SCIENTIFIC NATURAL PHILOSOPHY

By

E. E. Escultura

Research Professor

GVP – Professor V. Lakshmikantham Institute for Advanced Studies and Department of Physics, GVP College of Engineering, JNT University, Madurawada Visakhapatnam, AP, India

http://users.tpg.com.au/pidro/ **http://edgareescultura.wordpress.com/**

International Federation of Nonlinear Analysts

DEDICATION

This book is lovingly dedicated to my beloved wife, Violeta, who supported my academic endeavor all the way

CONTENTS

FOREWORD

It is astonishing how wide the enumeration of scientific fields is in this book. One can find here foundations of new mathematics with the new real number system; philosophical foundations of the new physics; theory of evolution; qualitative modeling and quantum gravity; cosmology and macro gravity; theory of intelligence and the nature of thought; geological and atmospheric turbulence and others. In each area of science the author has proposed new nonstandard approaches to its development and solved long-standing problems. The author has formulated about 50 laws of nature that cover vast areas of natural science.

His Grand Unified Theory (GUT) must find a deserving place among other theories claiming to be named Unified Theories. GUT as well as its theoretical applications is based on about 50 natural laws. Foremost among them are: the existence of two fundamental states of matter – visible and dark – and the basic constituent of matter, the superstring, that comprises both fundamental states. Suitable agitation converts dark or non-agitated and semi-agitated superstrings to agitated superstrings or visible matter. The author's attempt to lift the veil over dark matter deserves attention.

The author's important contributions in the new mathematics are the resolution of Fermat's Last Theorem (FLT); the solution of the gravitational n-body problem using the integrated Pontrjagin maximum principle; and the construction of the new real number system (new reals) including the dark number that qualitatively models the superstring.

There are many other new scientific approaches in the book. However, no one on Earth possesses Truth at the highest level and this book while proposing new approaches calls the reader to further investigation.

V. Gudkov

Professor
Institute of Mathematics and Computer Science
University of Latvia, Rainis Boulevard, 29, Riga, LV-1459, Latvia

PREFACE

Scientific natural philosophy is distinguished from previous philosophies for being based on the grand unified theory (GUT). Its methodology of qualitative modeling *explains* how nature works, while quantitative modeling *describes* nature's appearances mathematically. Therefore, GUT can be understood only as self-contained unified physical theory based on the laws of nature. However, the two methodologies are complementary, each indispensable to the other.

Chapter 1 deals with the foundations of mathematics and physics. Chapter 2 surveys the classical and new mathematics involved in the development of GUT but provides full details on the new real number system. Chapter 3 is the formulation of GUT through its three pillars – quantum and macro gravity and thermodynamics. Chapter 4 offers a wide range of applications of GUT from the theory of intelligence and the Earth sciences through astronomy, cosmology, biology, physical psychology, genetic engineering for the treatment of genetic diseases and optimal control theory through the integrated Pontrjagin maximum principle in the derivation of the quantitative component of the solution of the n-body problem. Chapter 5, the final and main chapter, provides the philosophical integration and summation of this work. Except for Chapter 2 the book is accessible to the general readership.

The author acknowledges with deep appreciation the contributions of colleagues, Professors C. G. Jesudason of the University of Malaya, for sustained debate on GUT, E. de la Cruz of California State University, Northridge, for constructive criticisms, and V. Gudkov of the University of Latvia for continued support and collaboration, all of whom contributed immensely to this book. Most of all, greatly indebted to Professor V. Lakshmikantham for paving the way to the frontiers of mathematics and science.

E. E. Escultura

Research Professor
GVP – V. Lakshmikantham Institute for Advanced Studies
GVP College of Engineering, JNT University, Madurawada Visakhapatnam 530041 AP
India

GENERAL INTRODUCTION

The novelty of our methodology calls for this general introduction to initiate the reader to the grand unified theory (GUT), the core of scientific natural philosophy. The conventional methodology of physics, quantitative modeling, has left long-standing problems unsolved, e.g., the turbulence and gravitational n-body problems, and fundamental questions unresolved, e.g., what the basic constituent of matter and the structure of the electron are, an inadequacy that prompted the author to introduce an alternative methodology – qualitative modeling – that digs deeper into nature beyond appearances to find out how it works. It explains not only the appearances of nature revealed by natural phenomena but also their dynamics including natural forces, interactions and behavior in terms of the laws of nature. Natural laws are discovered by observing patterns and regularity in nature and articulating them as natural laws upon which a physical theory such as GUT is built to express scientific knowledge as a deductive system subject to the most updated standards of mathematical rigor and precision. Thus, qualitative modeling gave birth to theoretical physics where there was only mathematical physics. The validity of a physical theory rests on its ability to explain natural phenomena and make verifiable predictions including invention of technology that works. Anytime a contradiction arises in a physical theory within its own structure or from experimental results it goes down the drain unless it is fixed by suitable modification or discovery of natural law that resolves it.

Just as all concepts of a mathematical space are defined by its axioms and conclusions derived from them, all physical concepts and structures, properties and interactions are determined and defined by the laws of nature and conclusions and predictions derived from them. For example, the structure and properties of the superstring are derived from natural laws. In other words, this new methodology axiomatizes physics as a deductive system where the axioms are laws of nature. In effect, it alters the primary task of the physicist from computation and measurement to the search for the laws of nature.

The boundaries between dark and visible matter and semi-and non-agitated superstrings are based on the finest arc length of visible light which may be refined in the future but our analysis will prevail as long as it is based on the laws of nature.

CHAPTER 1

Philosophical Foundations of Mathematics and Physics

Abstract: This chapter stands on the major rectification of foundations by David Hilbert in the early 20th Century and more. Recognizing the ambiguity of an individual thought being inaccessible to others Hilbert proposed that mathematics should be concerned not only with individual thought but with its representation by objects that can be studied collectively, including symbols, subject to consistent axioms. We stand on Hilbert's rectification and avoid other sources of ambiguity, where contradiction usually hides, such as infinity, large and small numbers and vacuous concepts and propositions. Turning on the real number system we find the field axioms inconsistent and rectification paves the way for its reconstruction into the new real number system under new set of axioms.

In physics we wonder why there are long-standing unsolved problems such as the gravitational n-body and turbulence problems and unanswered fundamental questions like what are the basic constituents of matter and the structure of the electrons. We conclude that its present methodology of quantitative modeling that describes nature's appearances mathematically is quite inadequate, and proceed to complement it with qualitative modeling that explains not only its appearances but also how it works in terms of its laws. Either methodology is indispensable to the other. Using qualitative modeling we proceed to participate in the 5,000-year search for the basic constituent of matter and succeed in pinning it down to the last detail of its structure called generalized nested fractal sequence. The superstring was the crucial factor in the solution of the 200-year-old gravitational n-body problem.

INTRODUCTION

We pose the problems and fundamental questions of mathematics and physics to guide the development of later chapters which will provide solutions and answers. We also summarize the critique-rectification of the foundations of mathematics that paved the way for the development of the mathematics of GUT.

THE UNSOLVED PROBLEMS OF MATHEMATICS

Mathematics has a number of unsolved problems but we focus on the most famous one, the 360-year-old conjecture called Fermat's last theorem [17, 118] that says,

For n > 2, the equation,

$$x^n + y^n = z^n,$$ (1)

has no solution in positive integers.

The conjecture appears simple and clearly stated but on closer look it is actually vague and it took the critique of its underlying fields – foundations, number theory and the real number system – to realize it. Then rectification was carried out to remedy the ambiguity and resolve the problem. For example, Peano's postulates that define the natural numbers are inadequate; number theorists do not even bother with them and use instead the real number system.

CRITIQUE-RECTIFICATION OF FOUNDATIONS, THE REAL NUMBER SYSTEM

We do not bother with the natural numbers because once the real numbers are fixed as the new real number system with the decimals as subspace and base we establish their isomorphism with the integers, i.e., the integral parts of decimals. We build on the great contribution of David Hilbert – the recognition of the ambiguity of the concepts of individual thought being inaccessible to others and, therefore, can neither be studied collectively nor axiomatized nor the subject matter of mathematics. For precision and clarity, the subject matter of mathematics can only be objects in the real world that everyone can look at and study such as symbols and material objects provided they are defined by consistent premises or axioms that specify their existence, behavior, properties and relationship among them. Then we call such system of objects and its axioms mathematical space. Thus, the game of chess is a mathematical space where the axioms are its rules. Tank and naval warfare are mathematical spaces where the

4 Scientific Natural Philosophy

appropriate principles of $\mathbf{R}^3$ and nature of weapons, tanks and battle ships are the axioms. In fact, tank battles inspired the fields of mathematics called game theory and operations research. In arithmetic, some properties of the decimals are taken as axioms. Without consistency a mathematical system collapses since any conclusion from it can be contradicted by another. The choice of the axioms is arbitrary depending on what the mathematician wants his mathematical space to do but once chosen, they become binding, i.e., every conclusion must follow from and every construction be justified by them to make it a deductive system.

Axiomatization objectifies a mathematical system and rids it of subjectivism by the mathematician as everything rests on the axioms including the rules of inference. The axioms well-define the mathematical system and its concepts completely and leave no room for universal rules of inference like formal logic since the latter has nothing to do with them. In this regard, all proofs of theorems involving mappings between distinct mathematical spaces are flawed. This applies to Gödel's incompleteness theorems [54 – 61].

This clarification is important as it puts an end to the confusion regarding the equation 1 = 0.99… that has generated much debate online during the last 13 years. These objects are distinct like *apple* and *orange* and to say that apple = orange certainly makes no sense.

The problem that confronted Hilbert was how to insure the consistency of a mathematical system. He proposed a consistent physical or mathematical model, i.e., an isomorphism between the model and the mathematical system. A physical model is, of course, consistent since its behavior is subject to the laws of nature which are consistent, otherwise, our universe would have collapsed a long time ago. It did not and has existed and evolved to higher order since its birth 8 billion years ago [42], e.g., in our young universe there were neither biological species nor biological laws that we now enjoy. To establish the isomorphism the binary operations of the mathematical system must have counterpart binary operations in the physical model. However, since our universe (distinct from the timeless and boundless Universe of dark matter [25] where our visible universe is a local bubble along with other universes) is finite and discrete, only a finite model is possible and the mathematical systems we can check for consistency are limited to finite systems which are quite inadequate for the purposes of science. The simplest and most developed mathematical system during Hilbert's time was arithmetic.

However, the incompleteness of arithmetic makes it ambiguous and unsuitable for modeling mathematical systems since ambiguity often hides inconsistency. The incompleteness of arithmetic means the existence of true proposition that has no proof. The absence of proof is due to the ambiguity of concepts, i.e., ill-defined by its axioms, and the present axioms of arithmetic (field axioms) are inconsistent [93]. Therefore, we look elsewhere to insure consistency of mathematical systems by identifying and avoiding the sources of ambiguity and contradiction or minimizing their impact. They are:

1) Large and small numbers due to our limited capability to compute their digits even with the most advanced technology.

2) Vacuous concept. For example, the concept "the root of the equation $x^2 + 1 = 0$ in the set of real numbers" is vacuous because the equation has no root and yet the concept $i = \sqrt{(-1)}$ is presented as its root which does not exist. Consequently, it yields these contradictions:

$$i = \sqrt{(-1)} = \sqrt{(1/-1)} = 1/\sqrt{(-1)} = 1/i = -i, \tag{2}$$

from which follows, 1 = 0, i = 0 and, for any real number r, r = 0, and both the real and complex number systems collapse. The remedy for the complex plane is in the appendix to [19]. Another vacuous concept is "the largest integer N". By the trichotomy axiom one and only one of the following holds: N < 1, N = 1, N > 1. The first inequality is out and if N > 1, then $N^2 > N$, contradicting the definition of N as the largest integer. Therefore, N = 1. This is the original version of the Perron paradox [117, 118]. (The trichotomy axiom is false in the real number system but follows from the lexicographic linear ordering of the consistent new real number system [21])

3) Self-referent or circular proposition where the conclusion applies or refers to the hypothesis. All the Russell paradoxes belong to this type; so does the indirect proof. We cite a few examples to illustrate the problems it gives rise to and find a remedy if possible.

a)   (Bertrand Russell) Let M be the set of all sets where each element does not belong to itself, i.e. M = {m: m ∉ m}. Then, either M ∈ M or M ∉ M. If M ∈ M, the defining conditions for M holds and M ∉ M. On the other hand, if M ∉ M, then M satisfies the defining condition; therefore M ∈ M. Self-reference follows from the fact that a set is defined by its elements and each element is defined by its membership in the set. It is a vicious cycle [78].

b)   The famous Russell antimony: A Cretan (native to Crete) saying "All Cretans are liars." Is he telling the truth?

c)   The barber paradox: The barber of Seville shaves those and only those who do not shave themselves. Who shaves the barber?

Russell's prescription to avoid this kind of difficulty is simply to keep away from it. The contradiction in (b) can be avoided by inserting the disclaimer "except me" after the phrase, "All Cretans". Similarly, the problem in (c) can be resolved by inserting "except himself" after the phrase "The barber of Seville".

4)   An infinite set is ambiguous since we do not know all its elements. Consequently, any statement involving the universal or existential quantifier on infinite set is ambiguous. Such statement is unverifiable and any definite or categorical statement about an ambiguous concept is also ambiguous. For example, to verify that every element of an infinite set has property A, we check an element to see if it has the property and keep checking "every" element. Obviously, we cannot exhaust all of its elements and thus we cannot verify if the property A holds for all elements of the set. The same problem is true of the existential quantifier. We cannot exhaust the elements of an infinite set to check that there is indeed such element with the specified property.

In fact, Lakatos' counterexample to every phase in Cauchy's proof of Euler's formula relating the edges, vertices and faces of a polyhedron in $\mathbf{R}^3$ [77] can be attributed to the fact that the set of such polyhedra is infinite and some exception hides behind each claim in the proof.

In mathematics, particularly theory of numbers, there are many statements involving infinite set of natural numbers that raise questions still unanswered, i.e., the statements are neither proved nor disproved. They usually stem from the inherent ambiguity of infinite set. Consider the following examples taken from [78]:

a)   A perfect number has the sum of its proper factors equal to the number itself. The first few known perfect numbers are 6, 28, 496, 8128, and 33, 550, 336 [78]. Question: are all perfect numbers even?

b)   Twin primes are prime numbers that differ by 2, like 3 and 5 or 11 and 13. The question is: are there arbitrarily large twin primes? Does there exist an infinite number of twin primes?

c)   Goldbach's conjecture [9] that says every even number except 2, is the sum of two primes. For example 4=2+2, 20 = 13 + 7, 30 = 19 + 11, etc. Question: is the conjecture true? This conjecture has been proved recently [23]. The uncertainty in the proof arising from the ambiguity of infinite set can be avoided in this theorem by re-stating it as follows: given any even number N except 2 there exist two primes whose sum is N. This theorem holds for any *given* even number N which is finite.

Consider the statement with the existential quantifier: The decimal expansion of a number p has no row of one hundred threes. True or not, it is unknown although extensive calculation on its decimal expansion has not yielded such row of threes. The probability that this statement is true is $1 - (9/10)^{100}$ which is almost 1. Thus, even a statement with probability that it is true is near 1 is not certain.

Among the field axioms that define the real number system $\mathbf{R}$ are (a) the trichotomy and (b) completeness axioms. The trichotomy axiom says that for any a, b ∈ $\mathbf{R}$, exactly one of the following is true, a = b, a < b or a > b; the completeness axiom says: every non empty subset S of R that is bounded above (has an upper bound) has the least upper bound.

The completeness axiom is a variant of the axiom of choice one version of which says, essentially, that if a soft ball is suitably sliced into infinitely small little pieces, then the pieces can be suitably rearranged, without distortion, and reconstructed into a ball, the size of the Earth (Banach-Tarski paradox) [75]. This is a contradiction in R³ inherited from the reals and attributed to the axiom of choice. Actually the axiom of choice is incidental here. The specific source of the problem aside from the ambiguity of infinite set is the Archimedean property of the reals that says: given any real number ε > 0 no matter how small and any number M, no matter how large, there exists some number N such that Nε > M. This allows one to form an arbitrarily large object from arbitrarily large number of arbitrarily small pieces [15, 75, 78].

A paradox is really a contradiction but others look at it as something unexpected or contrary to intuition.

A counterexample to any of the axioms of a mathematical system reveals inconsistency. Lately, L. E. J. Brouwer constructed a counterexample to the trichotomy axiom [5]. We present our version of the counterexample taken from [21] that shows at the same time that the real number system is not linearly ordered by "<" and that an irrational is not the limit of a sequence of rationals in the standard norm. (In fact, the notion *irrational* is ambiguous [21])

Let C be irrational. We want to isolate C in an interval such that all the decimals to the left of C are less than C and all decimals to the right of C are greater than C. We do this by constructing a sequence of smaller and smaller rational intervals (rational endpoints) such that each interval in the sequence is inside the preceding (this is called nested sequence of intervals). In the construction we skip the rationals that do not satisfy the above condition. Although given two distinct rationals x, y we can tell if x < y or x > y, we cannot line them up on the real line under the relation "<"since if x, y are two rationals, x < y, there is an infinity of fractions or rationals between them and we cannot verify their arrangement. (Not all reduced fractions or rationals are well defined, e.g., when the denominator of a reduced fraction is a prime other than 2 or 5, since we do not know all the digits of the quotient; in other words, there are missing elements among the fractions, the reason for the difficulty in arranging the decimals linearly with respect to "<")

Therefore, we settle for this scenario: starting with the rational interval [A, B] we find a nested sequence of rational intervals that "insures" C lies between the two endpoints at each stage. We go for an arrangement that will allow us to distinguish the left from the right endpoints of the sequence. We construct rows of rationals starting with numerator 1 in the first row, 2 in the second row, etc., and the denominators in each case consisting of consecutive integers starting from 1 in increasing order going right so that in each case we start with a denominator of a potential right endpoint.

Actually, we can squeeze the rows into a single row since no particular order with respect to "<" is involved. Even this arrangement is a problem. For example, suppose at a certain stage in the construction we have a left endpoint 1/5 then the number 20/100 appears on the right and, in trying to pair the left endpoint 1/5 with a right endpoint we skip 20/100 and all other rationals to the right of 1/5 in the ordering "<" that appear on the right and move further left than all of them. We choose the left endpoint in the succeeding step similarly to the right of 1/5, etc. Without loss of generality, we take this rational 1/5 to be the first left endpoint in the construction. Then once we have found the right pair for 1/5 we either use it as the right endpoint of the next rational interval and pair it with some rational on the right of 1/5 or find a new left endpoint to the right of the first left endpoint to pair with a right endpoint left of 1/5, etc. We make sure that we do not get closer to C than 10^{-n} at the nth step in the choice of the first n endpoints so that C remains inside each interval. While we are sure for all left and right endpoints A, B we have already identified in our construction, that A < C < B and all rationals right of A and left of B in the ordering "<" satisfy this inequality, there remains an interval of rational endpoints containing C and rationals that do not satisfy this inequality no matter how large we choose n. Therefore, the location of C remains unknown.

(In Brouwer's version of the counterexample, there is no limitation on how close the rational end points are to the irrational C. Therefore, by skipping the rationals that do not belong to the left or right endpoints, the right endpoints of the sequence in the construction eventually appear on the left and the left endpoints appear on the right, the irrational C nowhere to be found [5]. This was the first counterexample that showed the field axioms inconsistent and invalidated Wiles' proof of FLT [110, 111])

This construction attests to the ambiguity of the concept *irrational* and the problem of representing it as limit of a sequence of rationals; for every such sequence there is always a gap. Even the rationals in the real line are ambiguous mainly because there is an infinity of rationals between any given two rationals so that we cannot order them under the relation "<", i.e., we cannot line them up in the line interval between 0 and 1, denoted by [0, 1], under this relation. This is due to the ambiguity of infinity. Consequently, the real number system is not linearly ordered by this relation and the trichotomy axiom that says, given two real numbers x, y, one and only one of the following holds: x < y, x = y, x > y, is false. It shows further that the fractions are just as ambiguous as the nonterminating decimals the latter having a bit of advantage for being linearly ordered by the lexicographic ordering. We shall see that the new real numbers are linearly ordered by the lexicographic ordering "<" from which follows the trichotomy axiom.

We state the following two theorems; the proofs are standard and presented in Chapter 2.

Theorem

The rationals and irrationals are separated, i.e., they are not dense in their union (this is the first indication of discreteness of the decimals).

Theorem

The largest and smallest elements of the open interval (0, 1) are 0.99… and 1 − 0.99…, respectively [34].

QUALITATIVE MATHEMATICS AND MODELING

Some mathematical or scientific problems including proving or disproving Fermat's last theorem require more than just computation and measurement to resolve. For example, in an inverse problem in differential equations the boundary conditions are generally unknown or incomplete and what is known is their outcome. An example of an inverse problem is the gravitational n-body problem [41] posed by Simon Marquis de Laplace at the turn of the 17th Century that says:

Given n bodies in the cosmos at some initial time with their respective masses, positions and velocities and subject to their mutual gravitational attractions find their positions, velocities and paths at later time.

The bodies have history and boundary conditions way back in the past that gave rise to their present phase as a physical system. To solve this problem we must know what they were and the nature of the bodies and their interactions. Incidentally, it was this problem and the search for solution that gave birth to the grand unified theory [42].

Mathematical analysis such as the critique-rectification of foundations and the real number system cannot be done by computation or measurement alone. Therefore, we expand the tools of mathematics and science. Consider the following mental activity:

Making conclusions, visualizing, abstracting, thought experimenting, learning, doing creative activity, intuition, imagination and trial and error to sift out what is appropriate, negating what is known to gain insights into the unknown, altering premises to draw out new conclusions, thinking backwards, finding premises for a mathematical space and devising techniques that yield results.

This activity falls under rational thought or intelligence; we call its representation in the real world qualitative or non-quantitative mathematics. A lot of imagination as component of qualitative mathematics is needed to understand this book.

Since individual thought is not accessible to others and not all mental activity can be represented in the real world, there is inherent ambiguity in individual thought and, therefore, only its representation in the real world that can be studied and analyzed collectively is the proper subject matter of mathematics.

Qualitative mathematics introduced for the first time in and the main contribution of [46] includes abstract mathematical spaces, foundations, mathematical reasoning and, most important of all for our purposes, the search for the laws of nature. It is the main component of the new methodology of qualitative modeling that explains nature in terms of its laws applied to physics for the first time in [41] to solve the n-body problem. It provides the remedy for the inadequacy of computation and measurements that has left longstanding problems unsolved and fundamental questions unanswered. It alters the task of the scientist from computation and measurement that describe natural phenomena to the search for the laws of nature upon which physical theory is built to explain natural phenomena and solve scientific problems. Moreover, it raises the quality of scientific knowledge from description of the appearances of nature to a deeper understanding of how nature works by explaining its internal dynamics and the forces and interactions of physical systems in terms of natural laws.

THE PLACE OF MATHEMATICS IN SCIENCE

Like any language, the subject matter of mathematics is not nature but itself; therefore, it is not a science whose subject matter is nature. It is a specialized language that suits the needs of science. It articulates a physical theory so that the latter can explain natural phenomena, provide solution to scientific problem, predict the future course of a dynamical system and serve as guide in designing technology and scientific experiment. Mathematics is the medium of thought for studying our universe including our physical self that is external to our thought. We can also use it to study the representation of thought just as we study the structure or usage of a language. Physical theory is the form by which we express or articulate knowledge of nature.

The traditional role of mathematics has been computation for purposes of describing natural phenomena. Even Einstein's vision of uniting gravity and electromagnetism was aimed at describing both natural phenomena by similar equations, i.e., the same forms except for some constants of nature. He succeeded in doing so for gravity and electromagnetism using the field equations of relativity and Maxwell's equations of electromagnetism, respectively, but only in a very weak sense being sought at the time: Maxwell's and the field equations have similar forms. Then quantum physicists joined in to try to extend the unification in the same sense to the weak and strong forces of physics. It did not prosper, however, as they got stuck in the search for the basic constituent of matter which is central to the description of quantum physics.

Qualitative mathematics came to physics in a dramatic way: it was instrumental in the discovery of the 11 natural laws for the solution of the gravitational n-body problem in 1997 and the discovery of the basic constituent of matter required for it [41]. They anchored the initial formulation of GUT called flux theory of gravitation [33].

As the main component of qualitative modeling qualitative mathematics extended the theoretical applications of GUT to the broad fields of natural science and their applications as far as engineering, medicine [30, 32], physical psychology [28, 29], mathematics-science education [29], theory of evolution [30, 31], geological and atmospheric sciences and oceanography [35, 38] and design of technology along with the discovery of relevant laws of nature. For example, at least two laws of nature were discovered to explain why the final flight of the Columbia Space Shuttle ended in disaster killing all seven flight crew members aboard [40]. That catastrophic failure in technology has not been explained by conventional science and the same problem of breach of the insulation panel recurred during the resumption of the program, the reason for its termination.

Qualitative modeling found applications in another new field – complex systems, i.e., physical or social systems or problems that cannot be analyzed or solved by computation and measurement alone [6, 47]. They include generalized fractal such as the configuration of the superstring and the problem of economic-industrial development of underdeveloped and developing countries [6, 47].

PHILOSOPHICAL FOUNDATIONS OF THE NEW PHYSICS

The new physics elaborated in [78] is hybrid between the grand unified theory (GUT) and its mathematics [25, 42, 78]. Its main content, however, is GUT the foundations of which are the solution of unsolved problems and answers to the fundamental questions of physics [78]. We introduce them here and discuss them later. The unsolved problems include the long-standing gravitational n-body and turbulence problems and the more recent one posed by Einstein in the 1920s, the unification of gravity and electromagnetism and the weak and strong forces of physics in terms of similar mathematical description of their appearances. Our solution of the last problem is much broader: the unification of the forces and interactions of nature by a single physical theory – GUT [42]. The unanswered questions are listed in [78] but we give priority here to what the basic constituent of matter is.

THE UNSOLVED PROBLEMS AND UNANSWERED QUESTIONS OF PHYSICS

A problem is famous because it is simply and clearly stated but defies solution for a long time. Consider the famous gravitational n-body problem posed by Simon Marquis de Laplace at the turn of the 18th Century in his book *Celestial Mechanics* that defied solution for two centuries. It says: given n bodies in the Cosmos at a specific initial time with known masses, positions and velocities subject to their mutual gravitational attraction find their positions, velocities and paths at later time. (It was assumed that masses of cosmological bodies do not change but they do [25, 83]) The problem seems adequately and clearly stated for we think we know what bodies and gravity are which is not true. We do not know how bodies behave; nor do we know the full impact of gravity on them. In other words, the problem is unsolvable as it is. What would it take to solve it? We have to know what a body consists of which leads to the 5, 000-year search for the basic constituent of matter. Then we need to know what gravity is. Newton's so-called law that says the gravitational force of attraction between two bodies of masses m_1 and m_2 *s* cm apart is given by $F = Gm_1m_2/s^2$, where G is a known cosmological constant [12], is a description of motion of two masses subject to their mutual gravitational attraction. It appears that the solution is a matter of computation but it is not. The questions of what matter consists of and what gravity is cannot be answered by describing the appearances of bodies under the influence of gravity.

The other long-standing problem of physics, so long no one knows who posed it, was the turbulence problem [35]. The problem is to find out what gives rise to it, e.g., typhoon and tornado. Of course, it is not as clearly stated as the n-body problem. One does not even know where to begin the reason the problem has defied solution for so long. Qualitative mathematics provided the key to its solution [35, 78].

In either case, the solution required qualitative modeling that led to the discovery of the basic constituent of matter and the initial 11 laws of nature of GUT [41]

There are unanswered fundamental questions in physics the most ancient one being what the basic constituent of matter is the answer to which is the key to understanding of basic physical concepts such as gravity, black hole and the structure of the electron and atom.

THE SEARCH FOR THE SUPERSTRING

It was the development of quantum physics in the1920s and the splitting of the atom in the 1940s that rekindled the search for the basic constituent of matter in the 1950s that was in limbo for over 5,000 years. During that decade particle physicists embarked on the search for the basic irreducible elementary particles (prima) by smashing the nucleus of the atom to determine its constituents (dark matter was unheard of then). They started with the cyclotron, then the bevatron, linear accelerator, hadron collider and now the large hadron collider or CERN, a circular accelerator 7 km across that straddles between the boundaries of France and Switzerland [12] – all aimed at smashing the nucleus of the atom by an energized proton to find out what is there, specifically, to find the *true* or irreducible elementary particles, i.e., the basic irreducible elementary particles that comprise every atom. By the 1990s the search for the basic irreducible elementary particles was a complete success with the discovery of the +quark (up quark) and –quark (down quark) and the electron by J. J. Thompson in 1897 [12, 43]. They are basic because they comprise the light isotope of every atom; a heavy isotope has, in addition, the neutrino in the neutron [43]. We shall discuss them in detail later. Whatever particle physicists have achieved beyond this point is a bonus for natural science, a bonus for mankind.

The basic prima are produced at staggering rate in the inner core of cosmological vortices, in the vacuum of space and in the cellular membrane of living things.

CHAPTER 2

The Mathematics of the Grand Unified Theory

Abstract: This chapter surveys conventional and new mathematics involved in the development of GUT. The most important conventional mathematics includes the theories of generalized curves and surfaces and the integrated version of Pontrjagin maximum principle developed by L. C. Young. We summarize their original development here because they are quantitative models of many important physical concepts. The boundary year for the new mathematics is 1998, the year of publication of the counterexamples to Fermat's last theorem that proves the false conjecture and catalyzed the development of new mathematics such as (a) the new real number system, (b) generalized integral, derivative and fractal and (c) the complex vector plane (fully developed here except (c)). The introduction to the complex vector plane that rectifies the complex number system is presented but its full development requires re-writing of the rectified complex number system as the basis for rectification of complex analysis. The introduction of qualitative mathematics paved the way for qualitative modeling, the crucial factor for the discovery of the superstring and the 11 initial laws of nature required for the solution of the gravitational n-body problem by serving as foundations of GUT. The d-sequence of a dark number qualitatively models the superstring, the generalized curve quantitatively models its path and that of an elementary particle and the new real number system quantitatively models time and distance of ordinary space. The generalized surface quantitatively models the expanding Cosmic Sphere before its burst at $t = 1.5$ billion years from the start of the Big Bang.

INTRODUCTION

We survey the mathematics directly involved in GUT's development but provide details on the major ones: the new real number system and the generalized curves, integral, derivatives and fractal. We consider conventional mathematics of GUT developed before 1997 and call the rest new mathematics.

CONVENTIONAL MATHEMATICS

The most important conventional mathematics involved in the development of GUT are the theories of generalized curves and surfaces and the integrated Pontrjagin maximum principle the last one developed by L. C. Young in 1969 which he adapted to generalized curves of optimal control theory called relaxed trajectories [117]. It is an improvement over the original Pontrjagin maximum principle developed by Pontrjagin and his associates in 1962 [91].

Special Functions and Generalized Curves

We consider mischievous functions that render certain established methods ineffective. However, once tamed they are useful. Examples of mischievous functions are the infinitesimal zigzag and the wild oscillation,

$$\mathrm{Sin}^m 1/x, \ (\sin^n 1/x)(\cos x^m 1/x), \tag{1}$$

where m and n are integers.

The Infinitesimal Zigzag

We generate a sequence of functions (polygonal lines),

$$C_n: y_n = y_n(x), \ 0 \le x \le 1; \ n = 1, 2, \ldots, \tag{2}$$

that converges to curve C_0 point-wise (or in the sup norm) as follows:

Without loss of generality, take $C_1: y_1 = y_1(x) \ 0 \le x \le 1$, the polygonal line joining A and B formed by sides AD and DB of triangle ABD with vertices at A(0, 0), B(1, 0) and D(1/2, 1/2) (Fig. **1**). Note that their slopes are +1 and −1, respectively. For the second term in the sequence, join the midpoint P of AD to the midpoint Q of AB and point Q to the midpoint R of DB to form the polygonal line APQRB from A to B. We denote this function by $C_2: y_2 = y_2(x), 0 \le x \le 1$. We continue similar construction on the polygonal APQRB. From the geometry of the figure, the slope of

C_n, $n = 1, 2, \ldots$, at any point in [0, 1] is +1 or −1 except at corner points where it is simultaneously +1 and −1 (derivative does not exist in this set of measure 0 being countable). Also, the length $|C_2|$ of $C_2 = |C_1|$ = length of $C_1 = \sqrt{2}$. Continuing similar construction on the finer polygonal lines we obtain a sequence of polygonal lines C_n: $y_n = y_n(x)$, $n = 1, 2, \ldots$, $0 \le x \le 1$, having the following properties:

(a) $|C_n| = \int_\Delta ((1 + (y')^2)^{1/2} dx + \int_\Sigma ((1 + (y')^2))^{1/2} dx = \sqrt{2} = |C_1|$ **(3)**

where y_n' is the derivative of y_n; $\Delta = \{x \mid y_n'(x) = +1\}$ and $\Sigma = \{x \mid y_n'(x) = -1\}$ (the set at which $y_n' = \pm 1$ has measure 0).

(b) The sequence C_n: $n = 1, 2, \ldots$, is uniformly convergent point-wise and since each C_n is continuous the limit, C_0: $y_0 = 0$, $0 \le x \le 1$, is continuous. In fact, y_0 coincides with $y(x) = 0$, $x \in [0, 1]$, which is absolutely continuous. Hence, y_0 is also absolutely continuous.

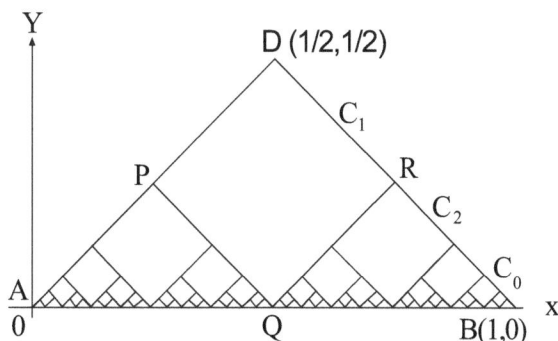

Figure 1: The first two terms of the sequence of polygonal lines that tends to the infinitesimal zigzag in the interval [0, 1].

(c) What about the derivative of y_0? Does it exist? If it does, what is it? We cannot have $y_0' = 0$. For, if that were so, it would violate the dominated convergence theorem applied to the integrals in (a) since this would imply,

$\int_{[0,1]} ((1 + (y_0')^2))^{1/2} dx = 1 \ne \sqrt{2} = \lim |C_n| = \lim \int_{[0,1]} ((1 + (y_n')^2))^{1/2} dx,$ **(4)**

as $n \to \infty$. Derivative of y_0 does not exist in the ordinary sense because the sequence y_n', $n + 1, 2, \ldots$, or −1, +1, −1 +1, …, does not converge to a single point since its set limit is $\{-1, +1\}$, *i.e.*, y_0' is set-valued.

The function C_0:(y_0, y_0'): $y_0 = 0$, $y_0' = \pm 1$, $0 \le x \le 1$ is a counterexample to a theorem in [93] that says, an absolutely continuous function is differentiable, almost everywhere. C_0 is absolutely continuous but nowhere differentiable. It raises some important points which are the source of this particular contradiction:

(a)   Inadequacy of the present notion of function; this was pointed out in [112, 113] 70 years ago and, again, more recently in [117]. A function defined by its values alone cannot distinguish the function C: $y = 0$ from C_0:$y_0 = 0$ which are distinct in at least two ways: one is differentiable and the other is not and they also have different lengths.

(b)   Inadequacy of the notion *derivative*; that the derivative of a function cannot be adequately expressed by its values because derivative is a property belonging to an extension of its underlying space (extension of n-space in the general case) whose restriction to the space of real-valued functions contradicts some of its properties (e.g., properties of absolutely continuous function). Therefore, there is a need to extend the notion of function to include those with set-valued derivative. Also, the present defect in the notion of limit is passed on to other notions defined by limits including the derivative [117].

The function C_0:$y_0 = 0$, $y_0' = \pm 1$, $0 \le x \le 1$, belongs to a wider class of curves called generalized curves different from the ordinary curve C: $y = 0$, $0 \le x \le 1$. Yet their values coincide point-wise. Furthermore, their arc lengths differ; in fact, there are countably infinite functions of this kind. One can see that although the sequence of functions C_n, $n = 1, 2, \ldots$, converges to the segment AB point-wise, its standard limit, say, in the sup norm, is something else: the infinitesimal zigzag, C_0:$y_0 = 0$, $y_0' = \pm 1$. This example raises two very important points:

(1)   Fallacious proof of existence of a mathematical object by approximation or convergence as well as the erroneous use of numerical and algorithmic methods without existence theory (in fact, this flaw is a variant of vacuous statement).

(2)   The inadequacy of the values of a function in characterizing its derivative; thus, the present notion of derivative is inadequate to capture the complexity of the property of a function.

This inadequacy of the notion of function and derivative as well as numerical method without rigorous justification has far-reaching significance for all of analysis and beyond. In particular, any theorem on derivative inherits this problem. That is drawn out by a property of a mischievous function that we shall deal with later. The infinitesimal zigzag is our first example of a mischievous function. It serves as counterexample to a number of well-established theorems.

Significance of the Infinitesimal Zigzag

This example says that the sup norm and the metric induced by point-wise convergence are not the natural metric for purposes of optimization, especially, in the calculus of variations; moreover, the properties of a curve are not fully accounted for by its values or parametric representation. We must put into account the behavior of its derivative and, as remedy, if f(t), t ∈ [0, 1], is the parametric representation of a curve C we represent it by the pair C: (f, g) where g is the derivative of f. Then the natural metric for purposes of optimization is the Young measure which we shall present later or curvilinear integral of some objective function along it (which can be the cost function) [117]. If we represent that measure by the integral,

$$I(C) = \int_{[0,\,1]}(f(t), g(t))dt, \tag{5}$$

then I(C) is the Young measure of the curve C. When the integrand is 1, I(C) is called the length of the curve. Thus, a curve is a linear functional and curves of the same Young measure belong to the same equivalence class representing that linear functional. This makes functional analysis available to optimal control theory. In an optimal control problem the derivative g is the control parameter so that it is independent of f; in other words, the system is controlled by finite set of values of the derivative.

Another case of optimization where the "obvious" curve is not the optimal solution is this example: find the minimum of the integral,

$$\int_{[0,\,1]}((1 + x^2)(1 + ((x'^2 - 1)^2)^{100})dt \tag{6}$$

(where x' is derivative) from 0 to 1 among admissible functions x(t) subject to x(0) = x(1) = 0. The "obvious" optimal curve among conventional curves is x = 0, subject to x(0) = x(1) = 0 and the minimum is 2^{100}. However, by admitting infinitesimal zigzag, which is like the ordinary curve x = 0 but whose derivative is set-valued and concurrently takes the values +1 and −1, and attaching a probability weight 1/2 to each of these values, we obtain a minimum of 1. Thus, the conventional theory of curves yields incorrect solution of this variational problem.

Here, the infinitesimal zigzag is a generalized curve or, to be precise, this generalized curve is the equivalence class of curves of the same Young measure (with set-valued derivative) [117]. Incidentally, all four types of cosmic waves and the superstring are generalized curves because they have one thing in common: set-valued derivatives; so is the path of a primum (elementary particle) [119].

The Wild Oscillation Sin1/x

Our next mischievous function is the wild oscillation, F(x) = sin1/x. This is a special case of the more general mischievous function $\sin^m 1/x^k$, where k, m, are positive integers. It reveals a flaw in the Lebesgue theorem on the Riemann integral that says:

A bounded function is Riemann integrable if and only if its set of discontinuity has measure zero [93]. The bounded function F(x) = sin1/x whose only discontinuity is at x = 0 is not Riemann integrable in any neighborhood of the origin. Known proof of integrability of sin1/x involves construction of a Riemann integral outside an ε-neighborhood of x = 0, where ε > 0, which exists, and taking a sequence of such integral as ε → 0, which converges.

The limit of such a sequence, however, is not necessarily Riemann integrable, certainly, not $\sin 1/x$ because no Riemann sum of this function can be formed in any neighborhood of 0. The best we can say here is that one can construct a convergent sequence of Riemann integrals with some relation to the function $\sin 1/x$ in the same ε-neighborhood of x = 0 but its limit is something else. This is, in fact, a form of the Perron paradox on the use of necessary conditions without an existence theory [117, 118]. It also illustrates the same fallacy mentioned earlier in the proof of existence of some mathematical object by approximation or convergence as well as the use of an algorithmic solution of a problem the existence of solution of which is not establish.

In the development of the Henstock integral in [80] the function $\sin 1/x^2$ plays a central role. However, the theory is flawed by the inadequate notion *derivative*. While this function is shrunk to zero by the factor x^2 its derivative is not for it belongs to a higher space independent of the function. The function considered in [80] is $x^2 \sin 1/x^2$, $0 \le x \le 1$. It is asserted that its derivative F'(x) exists at x = 0 and F'(x) = 0 because at that point its one-sided derivative can be trivially computed since, using the ordinary definition of derivative, we have,

$$| \Delta F | / | \Delta x | \le | x^2 | / | x | = | x |, \tag{7}$$

so that $\lim | \Delta F | / | \Delta x |$, as $x \to 0^+$, exists.

The inequality follows from the fact that F(x) is bounded by its envelope $y = \pm x^2$. F(x) is continuously differentiable outside x = 0. In fact, we have, at $x \ne 0$,

$$F'(x) = 2x\sin 1/x^2 - (2/x)\cos 1/x^2, \tag{8}$$

and its graph is shown in Fig. **2**. where the term $2x\sin 1/x^2$ has been discarded since it vanishes as $x \to 0^+$. However, the term $(2/x)\cos 1/x^2$ oscillates rapidly along all values in the interval $(-\infty, \infty)$ as $x \to 0^+$ and does not converge. This is a particular kind of discontinuity, an example of what we shall call chaos. Moreover, this is another example of a derivative of a function that is independent of it.

Our final mischievous function is a function of the type,

$$(e^{1/z}/x^k)(\sin^m 1/x^2 + \cos^n 1/x^2) \text{ or } (e^{1/z} x^k)(\sin^m 1/x^2), \tag{9}$$

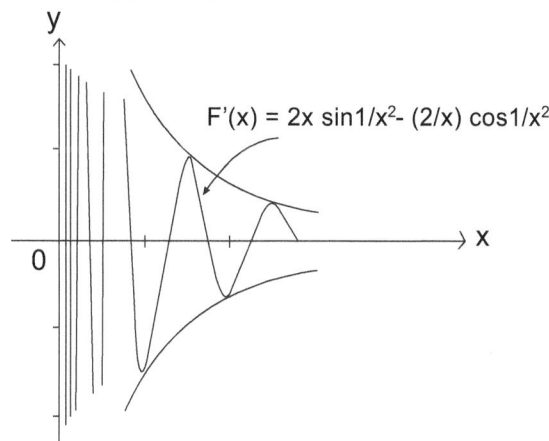

Figure 2: Graph of F'(x) = $-$ $(2/x)\cos 1/x^2$ where the term $2x\sin 1/x^2$ is discarded since it tends to 0 with x; it takes all values in $(-\infty, +\infty)$ as $x \to 0$.

where $z = x^2$, k, m, n are positive integers. Finding the limits of these functions, as $x \to 0$, quickly reveals that L'Hospital's rule breaks down on (9). The reason: these functions do not satisfy its hypothesis at the origin, that the function should not have a zero in any neighborhood; each of the functions in (9) has countably infinite zeros in any neighborhood of the origin. Also by rearranging the factors one gets different standard limits. The generalized derivatives of $(e^{1/z}/x^k)\sin^m 1/x^2$ and $(e^{1/z}/x^k)(\sin^m 1/x^2)$ or their expectations are evaluated in [2] and used to generalize L'Hospital's rule; the latter is applied to functions (9) to evaluate their limits as $x \to 0$ [2].

Rapid Spiral and Oscillation

A primum (unit of visible matter) is mathematically modelled by the rapid spiral $x = t$, $r(t) = \beta(\sin n\pi t)(\cos^m k\pi t)$, $t \in$ [$-1/k$, $1/k$], $\theta = nt$, n, m, k, integers, $n \gg k$, m even, whose profile is a sinusoidal curve of even power [25]. Its cycle energy is Planck's constant $h = 6.64 \times 10^{-34}$ joules [12], the irreducible unit of energy that we shall discuss later. Energy conservation and flux compatibility pull the primal cycles together to form a set-valued function that requires the generalized integral [19] to do calculation on it because the ambiguity or uncertainty of large number induces uncertainty on such large number of primal cycles.

(Note: a function multiplied by an oscillation is an oscillation)

Rectification of Inadequacy *Function* and *Derivative*

Partial rectification of the inadequacy of the concepts *function* and *derivative* is done in [112, 113] by representing a vector function f with values in **R** by a pair (f, g), where g is the derivative of f that takes values in a field of vectors belonging to a separate space isometric to $\mathbf{R}^n$. This representation formalizes the independence of the derivative from function. (For full development of this idea and the requirements on f and g see [112, 113]; we provide a summary later). Further rectification is required by the reconstruction of the real numbers into the consistent new real number system.

Regarding the derivative component of the pair (f, g) Young went so far as admitting set-valued derivative with the introduction of chattering controls [117]; in the study of convex vector functions the notion of a set of landing hyperplanes at a point is admitted. This corresponds to a set of tangent lines for a real-valued function [117]).

The thrust of rectification focuses on the derivative component of the pair (f, g) in the representation of a function. It is this approach that led to the introduction of generalized curves which, in turn, established an existence theory and resolved the Perron paradox [117] in the calculus of variations, paving the way for the latter's modern formulation in optimal control theory. However, with the appearance of set-valued functions in the study of fractal and chaos this rectification effort still falls short; we need to allow set-valued function in the pair (f, g) as a way to capture the complexity of certain notions, particularly, *function* and *derivative*.

Note that the approach in [113] reflects the methodology of enrichment: enriching the space with new elements to achieve an existence theory or convergence.

Let $\{f_n(t), g_n(t)\}$, $n = 1, 2. \ldots$, $t \in$ [a, b] be a sequence of functions in the new sense, where each $f_n(t)$ is continuous and $g_n(t)$ measurable and well-defined almost everywhere. We suppose further that the end points [a_n, b_n] of the domains of definition of $f_n(t)$ and $g_n(t)$ tend, respectively, to 0 and 1 as $n \to \infty$. For our purposes here we require that $a_n \geq 0$, and each of $f_n(t)$ and $g_n(t)$ has common extension to some interval T containing both [0, 1] and the sequence of intervals [a_n, b_n], $n = 1, 2, \ldots$. We define the limit set of $\{f_n(t), g_n(t)\}$ as the pair (($\{f_{0,0}(t)\}$, $\{g_{0,0}(t)\}$)), where $\{f_{0,0}(t)\}$ = Slim$\{f_{n,k}(t)\}$ = the set of limit points of a diagonal element $\{f_n(t_n)\}$, as $n \to \infty$ and $t_k \to t$ and $\{g_{0,0}(t)\}$ = Slim$\{g_{0,0}(t)\}$= the set of limit points of a diagonal element $\{g_n(t_k)\}$, as $n \to \infty$ and $t_k \to t$. Since f_n is continuous its limit is independent of the sequence $\{t_k\}$; not so with g_n since we only require measurability. Thus, we distinguish the limit set of a sequence of pairs $\{f_n(t), g_n(t)\}$ by the particular sequence $\{t_k\}$, where $t_k \to t$. This is consistent with our observation above that there is infinity of coincident but distinct curves on the segment AB, each element being determined by the particular curve that tends towards it. (As we shall see later those curves are countably infinite) As special case, let $t_k \to t$ for all t in T. Then the closure of such sequence $\{f_n(t), g_n(t)\}$ under this convergence is called the space of generalized curves. The complete formulation of the theory of generalized curves as linear functionals is given in [113]; we summarize it below. Essentially, a generalized curve is one with set-valued [113].

Applications of the Infinitesimal Zigzag

We make references to the superstring although we shall take it up later since we want to introduce its mathematical models. A superstring is a nested fractal sequence of superstrings where the first term is a close helix; it has a flux called toroidal flux in its helical cycle which is a superstring traveling at 7×10^{22} cm/sec or 10^{12} times the speed of light [4]; the toroidal flux has a toroidal flux in its helical cycle which is a superstring traveling at the same speed, etc. The projection of a helix on the plane through its axis is sinusoidal or oscillatory curve, by the energy conservation equivalence [25].

Given any curve in the plane we can deform it into an oscillatory curve $y = \sin^{m}bx$ which is rectifiable; we can further deform it into some isosceles triangle ADB so that its length is preserved and equal to the sum of the lengths of AD and DB. In turn, we can deform this triangle into a finer oscillatory curve K_1, with length preserved (Fig. **3**). We iterate this deformation forming an alternate sequence of polygonal lines and oscillatory curves K_n from A to B. Again, the sequence K_n tends towards a generalized curve called infinitesimal oscillation whose function component coincides with the zero function C: $y = 0$, $0 \leq x \leq 1$. Its length is equal to the original length $|K|$ of K and its derivative at any point $x \in [0, 1]$ is set-valued and equals the set of limit points of the derivatives of the sequence of oscillations at x.

Since the segment AB is arbitrary we can prescribe its length to be an arbitrary number $\varepsilon > 0$. Then we have the following:

Theorem 1

Given an oscillatory curve K, any number $\varepsilon > 0$ and a line segment AB, there exists a continuous deformation of K into a fine oscillatory curve inside an ε-neighborhood of AB that preserves the length of K [24, 34].

Theorem 2

Given an oscillatory curve K, there exists a continuous deformation of K, with length preserved, into an arbitrarily small neighborhood of a point [24, 34].

Proof

We prove both theorems. Let A be a given point and B a point in the ε-neighborhood of A and suppose $|AB| = \varepsilon/2 > 0$. There exists a deformation of K, with length preserved, into two sides of an isosceles triangle ADB where $|AD| + |DB| = |K|$. Following the construction above there exists a sequence of polygonal curves C_n and corresponding oscillatory curves K_n such that for each n, $K_n = C_n = AD + DB = K$ and K_n tends to the segment AB (Fig. **3**). Hence there exists a positive integer N such that whenever $n \geq N$, the curve K_n lies inside the ε-neighborhood of A. (This establishes the first theorem) since the length of AB is arbitrary, $\varepsilon > 0$ and $\ulcorner AB \urcorner = \varepsilon/2$. Then the second theorem follows from the first).

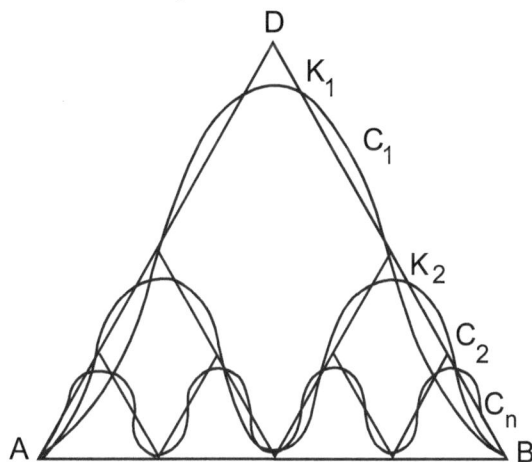

Figure 3: Sequence of sinusoidal curves of length $|K|$ that tends towards infinitesimal oscillation of the same length.

Note that in each case the oscillatory structure is preserved as well as its length. Thus, it is possible to shrink an oscillatory curve of any length into an infinitesimal oscillation at a point. Now, let $\beta > 0$, where β is small, and let K be an oscillatory curve of large length K. Let $\varepsilon = \beta/2 < |K|/2$. As before, we deform K into the two sides of an isosceles triangle ADB with base AB, where $\ulcorner$ AB$\urcorner$ = ε. Let h be the altitude of this triangle, then for suitably small ε, h ≈ | K | /2. By the

Archimedean property of the decimals there exists some positive integer n such that

$$|K|/2^{n+2} < |K|/2^{n+1} \leq |K|/2^{n}. \tag{10}$$

Therefore, in the sequence of oscillatory curves K_i with $|K_i| = |K|$, for each i = 1, 2,..., which tends towards the line segment AB, there is one whose amplitude satisfies the inequality (10). We state this as a theorem.

Theorem 3

Let k be an oscillatory curve with large length $|K|$ and let ε > 0, ε = β/2 < $|K|/2$. Then one can continuously deform the oscillatory curve K into an arbitrarily small neighborhood of a point with its length and amplitude prescribed to satisfy,

$$|K|/2^{n+2} < |K|/2^{n+1} \leq \varepsilon \leq |K|/2^{n}, \tag{11}$$

for some integer n [24, 34].

The following theorem [24] is now obvious and follows from the above theorems:

Theorem 4

The real line is chaos.

Theorems 1 – 4 model different aspects of the shrinking of a superstring. They have other implications for physics that can explain certain phenomena such as the tremendous but undetected (latent) energy in the nucleus of an atom. Tremendous because we can pack infinitesimal helical loops into an arbitrarily small neighborhood of a point at very high energy level hζ where h is Planck's constant and ζ is number of helical cycles. They cannot be detected by our means of observation such as light since if the wavelength of the latter is sufficiently fine there would be no interference or discordant resonance with this infinitesimal helix due to difference in orders of magnitude of its cycles (the helix can be semi- or non-agitated superstring [25]). Helix and oscillation (sinusoidal) are universal configurations of matter and they are related: the projection of a helix on a plane through its axis is sinusoidal. Infinitesimal helix and oscillation are both generalized curves because their derivatives are set-valued. The superstring, basic constituent of matter, is an infinitesimal helical loop, a generalized curve [25, 113].

The Generalized Curves

We summarize the development of the generalized curves [112, 113, 117] as an offshoot of the search for existence theory in the Calculus of Variations to resolve Perron paradox [117, 118] coming from the use of necessary condition without existence theory. We reconstruct its basic formulation to get a sense of the method of enrichment.

Consider the parametric equation x(t), $t_1 \geq t \geq t_2$, of the curve C, where x(t) is absolutely continuous and takes values in $\mathbf{R}^n$. At each t we take a vector y(t) taken from a vector field in a separate space isometric to $\mathbf{R}^n$ which we also denote by $\mathbf{R}^n$. We require the vector y(t) to lie in a unit sphere S, *i.e.*, $|y(t)| = 1$. We also require C to lie in a compact cube in $\mathbf{R}^n$. and denote by **A** the compact Cartesian product of these two sets. Our Lagrangian belongs to the space $C_0(A)$ of continuous functions with compact support **A**. We further assume that our Lagrangian is homogeneous in y so that it is determined in $\mathbf{R}^n$ by its restriction to **A** since if L is the Langrangian and (x, y) is any point in $\mathbf{R}^n \times \mathbf{R}^n$ then there exists some scalar α ≥ 0 such that y = αŷ, where ŷ∈ S. From the homogeneity of L in y,

$$L(x,y) = L(x,\alpha\hat{y}) = \alpha f(x,\dot{y}), \tag{12}$$

where f ∈ $C_0(A)$ is the restriction of L to **A**.

We make a minor adjustment for simplicity of notation by representing C(x(t), y(t)), a ≥ t ≥ b, as a curve defined on the compact set **A**. Here we have attached a derivative y(t) to stress the independence of derivative from x(t). We also assume the parameter t to be an arc length from the initial point of C to avoid dependence of the curvilinear integral along C on its parametrization but more on intrinsic properties.

We consider two curves C_1 $(x_1(t), y_1(t))$, $a \geq t \geq b$ and C_2 $(x_2(t), y_2(t))$, $c \geq t \geq d$, equivalent if their curvilinear integrals satisfy,

$$I(C_1) = \int_{[a,b]} f(x_1(t),y_1(t))dt = \int_{[c,d]} f(x_2(t),y_2(t))dt = I(C_2), \tag{13}$$

for all $f \in C_0(\mathbf{A})$. Thus, a curve C_0 is completely determined by the values of its curvilinear integrals all $f \in C_0(\mathbf{A})$, *i.e.*, a linear functional g in the dual space $C^*(\mathbf{A})$ defined on $C_0(\mathbf{A})$.

We define fine convergence of a sequence g_n, $n = 1, 2,...$, of elements of $C^*(\mathbf{A})$ and say that g_n converges to g_0 or $g_n \rightarrow g_0$ if $g_n f \rightarrow g_0 f$ for all $f \in C_0(\mathbf{A})$. The closure of this space of curves in the new sense is called the space of generalized curves. In this sense a generalized curve is the fine limit of ordinary curve.

A generalized curve is also called a generalized flow, where the latter is an element of the positive cone of $C^*(\mathbf{A})$, *i.e.*, $gf \geq 0$ for all $f \in C(\mathbf{A})$ with $f \geq 0$. It is the space of generalized curves that provides an existence theory in the calculus of variations and optimal control theory and renders Perron paradox inoperative there, showing again how a contradiction is resolved by some sort of enrichment, embedding or completion. This fine convergence induces a metric on $\mathbf{R}^n$ called Young metric. Thus, the distance between two curves C_1 and C_2 is given by,

$$I((C_1) - I(C_2)). \tag{14}$$

This norm is really the sup norm in the space of generalized curves defined as,

$$\sup |g_1 f - g_2 f| \text{ for all } f \in C_0(\mathbf{A}). \tag{15}$$

We require these linear functionals g_1, g_2 to be well-defined in the intervals of definitions of their respective duals (ordinary curves) in $\mathbf{A}$ and have common endpoints, the same requirement for fine convergence, that the dual sequence of g_n in $\mathbf{A}$ must have endpoints tending towards those of its limit g_0. The length of a generalized curve g is the value of the integral (15) when $f = 1$. This norm is the right one consistent with the counterexample presented earlier, the polygonal line that converges to an infinitesimal zigzag. In that case the function f is the arc length $((1 + (y_0'(t))^2)^{1/2}$ which is constant and equal to $\sqrt{2}$ for each C_n. Thus, $g_n f = \int_{[0, 1]}((1 + (y_0'(t))^2)^{1/2}dt = \sqrt{2}$. Of course, the derivative of f, $y_0'(t)$, is set-valued with value $\{1, -1\}$, and its dual in $C^*(\mathbf{A})$, where $\mathbf{A}$ $[0, 1] \times [0, -1]$, is the infinitesimal zigzag.

The development of the generalized curves started 70 years ago but the study of fractal bifurcation and chaos can benefit much from it, not to mention its many applications. In fact, a generalized curve such as the infinitesimal zigzag is both chaos and limit set of fractal but it needs to be rediscovered because contemporary studies on fractal have recent origin. In the case of bifurcation (or more generally, multifurcation), which is the transition from chaos to fractal, it can be explained by the fact that even well-behaved functions y = f(x) passing through some point (x_0, y_0) is only one of the countable infinity of local solutions of some differential equation near (x_0, y_0) satisfying an initial condition on its derivative there. Put another way, near the point (x_0, y_0), there is a countable infinity of functions (f(x), g(x)), which are local solutions of some set-valued differential equation. For each choice of g in a set-valued differential equation of the form x' $\in \{g(t, x(t)\}$ we have, for a given probability distribution, a corresponding branch of the solution. This is how multifurcation occurs at every point on the initial function as well as on each branch. This is how chaos ultimately results. This is similar to the formulation of the development of generalized surfaces that we used for dealing with the undecidable proposition FLT. Later, it was extended to relaxed trajectories of control theory [117].

(For extension of this methodology to the development of generalized surfaces see [50, 114 – 116])

To summarize, the generalized curve quantitatively models the superstring, primum and its path in flight [119] and spiral path of visible matter falling into the core and spinning around the eye of a cosmological vortex which are continuous arcs with set-valued derivatives [20, 78]. What we see in a spiral nebula are its stars and minor vortices falling into its core and spinning around the eye. Rapid spiral quantitatively models a simple primum and rapid oscillation the photon and primum in flight [25]. **R*** quantitatively models physical time and distance, non-standard g-sequence of d* the nested fractal superstring and d* the tail end of its toroidal fluxes, a superstring and continuum. The decimals model the metric system and the integers the countably infinite and discrete dark and visible matter; and, as we shall see later, GUT qualitatively models our universe. Note that a physical system may have more than one mathematical model but only one qualitative model within equivalence.

The Generalized Surfaces

We summarize Young's theory of generalized surfaces and his joint work with W. H. Fleming [50, 114 – 116]. We shall focus on the more developed version in [116] for $\mathbf{R}^m$, $m \geq 3$.

Let $\mathbf{X}$ be the Euclidean m-space and let the parameters (u, v) range in the unit square $R(0 \leq u \leq 1, 0 \leq v \leq 1)$. We say that x(u, v) is a generalized Dirichlet representation of if it is defined and bounded in R and takes values in $\mathbf{X}$, absolutely continuous on the intersection of R with almost every line u = constant or v = constant, its extension to the perimeter of R is continuous and its dirichlet integral,

$$L(x,R) = (1/2) \int_R ((x_u)^2 + (x_v)^2) dudv, \tag{16}$$

is finite.

We say that $x \in \mathbf{X}$ if x is an element of m-space, $x \in D$ and x is a generalized Dirichlet representation x(u, v). We also write $x \in D(N)$ to mean $x \in D$ and $L(x, R) < N$ for some positive integer N. The jacobian j(u, v) of x(u, v) is the vector product of the partial derivative x_u and x_v, *i.e.*, their normal, and exists almost everywhere in R. we write (x, j) $\in$ D to mean $x \in D$ and j is the jacobian of x. The values of any j(u, v) are bivectors. They lie in the space $\mathbf{J}$ whose elements can be identified with the skew-symmetric matrices j $=(j_{rs})$, r, s, 1, …, m, of rank 2 or 0. We define the norm $|j|$ for $j \in J$ as the square root of $\Sigma(j_{rs})^2$ summed up for $1 \leq r < s < m$). We denote by J_1 the subset of J consisting of the j of norm 1 and by J* the hyperspace obtained from J by deleting the condition that j be of rank 2 or 0. We write further (E, J), (E, J_1), (E.J*) for the Cartesian product of a subset E of X with J, J_1, J*.

We write for the space of continuous functions f = f(x, j) $\in$ (X, J_1) with norm $|f| = \sup | f(x, j) |$ for $x \leq B$, where B is sufficiently large constant depending on the context. Just as for generalized curves we require homogeneity of f in j, *i.e.*, f(x, αj) = αf(x, j) for $\alpha \geq 0$ so that the extension of f is defined by its restriction to F. the function f is called integrand and this is analogous to the Larangian for generalized curves. The classical generalized integral is defined as,

$$L(f) = \int_R f(x(u,v), j(u,v)) dudv, \; f \in F \tag{17}$$

A parametric surface is the equivalence or maximal class consisting of $x \in D$ all of which give the same value to the functional L(f) for each integrand f (or for its symmetric integrand). Each integrand f in the class is a representation of L(f). We define L(f) a generalized surface. A generalized surface L(f) is termed fine limit of the sequence of generalized surfaces $L_n(f)$, n = 1, 2, …, if for each relevant integrand f (each f in or symmetric f) the values of $L_n(f)$ converge weakly to L(f). It can be seen that every generalized sequence is expressible as the limit of a sequence of uniformly bounded parametric surfaces (ordinary surfaces). Such convergence induces a norm on the linear functional representing surfaces defined by the expression sup $|L(f) – L'(f))$, where the supremum is taken over all f $\in$ F whose norm is ≤ 1 (or symmetric f). When $f_0(x, j) = 1$ for (x, j) $\in$ (X, J_1) we term $L(f_0)$ the area of the generalized surface L. Note that in this formulation a generalized surface is a linear functional. The area of a generalized surface L(f) is L(1) (which corresponds to the length of a generalized curve gf when f is the constant 1).

The requirement of continuity for f is not necessary; measurability suffices. This can be further weakened by allowing set-valued integrand with appropriate probability distribution. In fact, in taking the limit of parametric representation of surfaces we may have set-valued integrand. Therefore, we introduce the notion of generalized derivative of an integrand as the expectation or average of its set-values at a point in each argument. This is the same as the equivalence class of probability distributions that yields the same expectation. For surfaces we can also admit set-valued jacobian with a probability distribution. This way we can deal with the notions of bifurcation and multifurcation of surfaces and chaos. Each surface that emerges has a certain probability of being actualized, such probability being determined by the probability distribution of its jacobian.

The space of generalized surfaces is complete in the norm defined by this fine convergence. This provides the needed existence theory which is local, *i.e.*, the existence of a surface subject to suitable initial and boundary conditions is valid only in a suitable neighborhood of a point. This is all we can expect since it is defined by differential equations describing local properties of the underlying space. However, it is shown in [114 – 116] that a global generalized surface is the fine limit of a sequence of suitable conventional surfaces. In Young's theory, the space of surfaces is enlarged to insure existence. This is achieved by allowing set-valued jacobians of a vector partial differential equation satisfying the requirements in [114 – 116] with unit measure or probability distribution. A generalized jacobian is the expectation of a set-valued jacobian with probability distribution. Thus, the local solution of a partial differential equation with set-valued jacobian depends on its probability distribution; consequently, the solution space forms a family of surfaces each of which is a representation of a generalized surface. A conventional surface is a solution of a conventional partial differential equation (with well-defined jacobian) satisfying requirements in [114 – 116]. Probability distribution introduces uncertainty in the space of generalized surfaces. Another level of uncertainty is brought in by the proof of local existence of a surface which uses fine convergence of a sequence of conventional surfaces along with the completeness of the space of generalized surfaces.

The Cosmic Sphere [42] which we shall discuss later is modeled quantitatively by a generalized surface.

The Integrated Pontrjagin Maximum Principle

We summarize the integrated version of the Pontrjagin maximum principle that we shall need in Chapter 4.

The Integrated Version

This version of the Pontrjagin maximum principle is developed in [117] (with slight improvement in [22]) from the original principle by Pontrjagin and his associates [91] to apply to generalized curves, specifically, relaxed trajectories of optimal control theory. Anchored on some existence theorems, it was meant to rid optimal control theory of the Perron paradox, a contradiction arising from the use of necessary condition without existence theory [117]. The principle itself is a necessary condition consisting of three parts but it rests on proof of existence of optimal solution of the optimal control problem among relaxed trajectories.

This principle was used to calculate the trajectories, positions and velocities of the n bodies in the gravitational n-body problem once the qualitative solution was provided by GUT [41]. But why do we really need to go through this level of sophistication when GUT has already determined that the n-bodies fall to or orbit around the cores of their respective cosmological vortices along rotating spiral streamlines? The streamlines are not ordinary curves but generalized curves, *i.e.*, they have set-valued derivatives, and for purposes of applications we need the specific equation of the trajectory of each body in the problem. In fact, we can look at a falling body as being subjected to two-valued control set one value being the pull of gravity and the other the impact of the centrifugal force of spin. The path of a body is what is called infinitesimal simplicial curve, *i.e.*, piece-wise arcs corresponding to alternating constant values of the control set U consisting of two elements. A generalized curve called relaxed trajectory is the limit of a sequence of simplicial curves in the Young measure [117]. Superposed on it is another generalized curve due to the micro component of turbulence but has no visible impact on this macro problem.

Formulation of the Problem

We first summarize the formulation of the time-optimal control pre-problem for each body in the original naïve optimal control theory [117], *i.e.*, without the benefit of existence theory. Then we update the formulation to be able to utilize the Integrated Pontrjagin Maximum Principle. We ask for the minimum of the integral,

$$\int f(t,x,u,w)dt, \qquad \qquad \textbf{(18)}$$

for trajectories x(t), controls u(t) and constants w and subject to,

$$\dot{x} = g(t,x,u,w)dt, \tag{19}$$

where u(t) ranges in the set U, w in W and subject to suitable specified conditions (dimensionality is at our disposal). We can eliminate the constants w by regarding the pair (x, w) as a point in higher space and adjoining the differential equation,

$$dw/dt = 0, \tag{20}$$

which insures that w is constant along any trajectory. Then we add to our end condition that the initial or final values of the projection w of (x, w) lie in W. Thus, the only effect of the constants is to alter dimensionality and end conditions.

To further simplify the problem we introduce another coordinate x_0 subject to,

$$\dot{x}_0 = f(t,x,u), \tag{21}$$

and write x and g for the pairs (x_0, x) and (f, g). We also add the end condition x_0 at the final end of a trajectory which correspond to the time t = 0; in other words, we reverse time and find the minimum of $-x_0$ for trajectories x(t) and u(t) subject to

$$\dot{x} = g(t,x,u), \tag{22}$$

where u(t) ranges in U and x(t) satisfies appropriate end conditions. Thus, in this problem what is being minimized is the function x_0 which in applications can be the cost function. Without loss of generality we assume that the pair of endpoints of x(t) belongs to some pre-assigned closed set B of the Cartesian product of x-space with itself (*i.e.*, the initial values of t is not restricted directly, e.g., we can set t + t_0 in place of t). We denote by G(t, x) the set of values of the vector g(t, x, u)

for fixed (t, x) as u ranges in U. Then we ask for the minimum of $-x_0$ subject to the condition,

$$\dot{x} \in G(t). \tag{23}$$

The problem with constraint (22) is called the controlled pre-problem; the one with constraint (23) the uncontrolled pre-problem. The latter has larger space of trajectories from which to find the minimum. As the space of trajectories becomes larger the better is the chance of existence of minimum. The space of trajectories of either problem is still conventional; we can still improve the solution by enlarging the space beyond conventional trajectories into the space of generalized curves. This is exactly what we need for the gravitational in-body problem because the trajectories we are looking for are the spiral streamlines each of which is the local resultant of the effect of gravity and centrifugal force mathematically modeled by relaxed trajectory. L. C. Young set the machinery for doing so [117] as follows.

Instead of the constraint equation,

$$\dot{x} = u, \tag{24}$$

where the velocity vector is controlled directly and coincides with the control vector u we attach a probability or unit measure to u (normalized probability distribution) so that the actual velocity dx/dt becomes the integral of u with respect to this unit measure. This is called the weighted average or expectation value for this probability measure. Then the control function u(t) that yields a specific trajectory be replaced by the probability measure v(t). Such measure is called chattering control value v, and we say that it reduces to a conventional control u if the measure is totally concentrated at u. We write V for the space of chattering control values v, *i.e.*, the set of unit measures on U.

Existence Theorems

We quote some existence theorems on conventional and relaxed trajectories from [117].

(1) Existence of Solutions to the General Initial Value Problem for Ordinary Differential Equations

Let f(t, x) be a vector-valued function with values in n-space and suppose in some neighborhood of (t_0, x_0), f(t, x) is continuous in x for each t, measurable in t for each x, and uniformly bounded in (t, x). Then there exists an absolutely continuous function x(t) defined in some neighborhood of t_0 such that $x(t_0) = x_0$ and, almost everywhere in that neighborhood,

$$\dot{x} = f(t, x(t)) \tag{25}$$

This theorem is of fundamental importance in both optimization and approximation theories, calculus of variations and optimal control.

(2) Halfway Principle of McShane and Warfield

Suppose given a continuous map p* from Q to P, and a measurable map p** from R to P such that

$$p^{**}(R) \subset p^*(Q) \subset P \tag{26}$$

Then there exists a measurable (lifting) map q* from R to Q such that

$$p^{**} = p^* q^*. \tag{27}$$

We denote by $\hat{G}(t, x)$ the set of the values of G(t, x, v) when (t, x) is kept fixed and v allowed to vary in V.

(3) First Corollary

Let x(t) be continuous in the finite time interval T and let z(t) be measurable vector-valued function in T such that $z(t) \in G(t, x)$ or $z(t) \in \hat{G}(t, x)$. Then there exists a measurable conventional or chattering control u(t) or v(t) such that z(t) = g(t, x(t), u(t)) or z(t) = g(t, x(t), v(t)).

(4) Second Corollary (the Filippov Lemma)

If, in particular, x(t) is an (uncontrolled) conventional trajectory subject to $\dot{x} \in G(t, x)$ or relaxed trajectory subject to $\dot{x} \in \hat{G}(t, x)$ almost everywhere. Then there exists a measurable conventional or chattering control u(t) or v(t).so that x(t) coincides with the corresponding controlled trajectory satisfying the differential equation

$$\dot{x} = g(t, x(t), u(t)) \text{ or } dx(t)/dt = g(t, x(t), v(t)), \tag{28}$$

almost everywhere.

(5) Uniqueness Theorem for the Initial Value Problem of an Ordinary Differential Equation dx/dt = f(t, x)

Suppose, in addition to the hypothesis of (1), that for some constant K, the function f(t, x) satisfies, whenever (t, x_1) and (t, x_2) lie in some neighborhood N of (t_0, x_0), the Lipschitz condition

$$/f(t, x_2) - f(t, x_1)/ \leq K / /x_2 - x_1/. \tag{29}$$

Then in some neighborhood of t_0 there exists one and only one absolutely continuous function x(t) such that

$$x(t) = x_0 + \int_\Delta f(\tau, x(\tau)) d\tau, \tag{30}$$

where $\Delta = [t, t_0]$.

Let $f \in C_0(T \times U)$, the space of continuous functions on f(t, u) on $T \times U$, where U is the set of control values u, and T is some fixed time interval. We write Δ for some variable time interval of U. For such pair (f, Δ) consider the function w of (f, Δ) defined by the integral

$$w(f,\Delta) = \int_\Delta f(t,v(t))dt, \tag{31}$$

where $v(t)$, $t \in T$ is a measurable chattering control (or v is conventional control, *i.e.*, unit measure concentrated at some point $u \in U$). We understand the integrand $f(t, v(t))$ as shorthand notation for an integral of f, for constant t, with respect to probability measure $v(t)$ on U. then we regard w as a measure and identify $w(f, \Delta)$ with the integral

$$\int_{\Delta \times U} f dw. \tag{32}$$

Then we write,

$$w = v(t)dt, \tag{33}$$

Bearing in mind that (31) is a double integral; thus, every control measure is determined by a chattering control $v(t)$.

The control measure w will be termed simplicial if it is defined by (31) where $v(t)$ reduces to a conventional piecewise constant control $u(t)$. Then we use formula (32) and reinterpret $u(t)$ as a unit measure on U concentrated at one point $u(t)$. We denote by W the space of all control measures w. We introduce fine convergence in this measure. A sequence of control measures w_v, $v = 1, 2, \ldots$ is termed convergent if, for each f, the values $w_v(f, \Delta)$ tend to a limit $w(f, \Delta)$ uniformly in Δ; then we say that w_v tends to w.

(6) Theorem

(i) The space W is sequentially complete. (ii) In order that a real function w of f and Δ be of the form w(f, Δ) where $w \in W$, the following system of conditions is both necessary and consistent.

  (a)   w(f, Δ) is linear in f and additive in Δ;

  (b)   f(t, u) ≥ 0 in $\Delta \times U$ implies w(f, Δ) ≥ 0;

  (c)   f(t, u) = 1 in $\Delta \times U$ implies w(f, Δ) = $|\Delta|$.

  (d)   Theorem. (i) The space W is sequentially compact. (ii) Simplicial control measures are dense in W.

We term bundle of relaxed, conventional or simplicial trajectories the family of trajectories which meet a given bounded closed subset of (t, x)-space corresponding to closed time intervals, possibly degenerate ones all contained in a fix time interval.

A sequence of functions

$$x_v, t \in T_v, t \in T_0, v = 1, 2, \ldots, \tag{34}$$

where T_v are closed time intervals all contained in some fixed time interval will be said to converge uniformly to

$$x_0, t \in T, \tag{35}$$

if, first, T_0 x_v, $t \in T_0$, a closed time interval whose extremities are the limits of those of T_v, and, second, for some choice of a closed time interval T containing T_0 and all but a finite number of the T_v, these exist, for large v, extensions of our functions of the form,

$$x_v, t \in T, \tag{36}$$

which tend uniformly to a corresponding extension to T of $x_0(t)$ (T_0 may be a point).

(7) Theorem

A bundle of relaxed trajectories is sequentially compact and complete and the corresponding bundle of simplicial trajectories is dense in it.

(8) Corollary

Suppose the set G(t, x) of the values of g(t, x, u) for fixed (t, x) is convex. Then any bundle of conventional trajectories is sequentially complete and compact, and the corresponding bundle of simplicial trajectories is dense in it.

(9) Existence Theorem for Relaxed Solutions

Let Q be a bounded closed set of (t, x)-space, P a closed set in the Cartesian product of (t, x)-space with itself and T a closed finite interval of t. We denote by Σ the set of relaxed trajectories x(t) defined on closed subintervals of T which meet Q and p a pair of extremities situated in P. The function g(t, x, u) which appears in the differential equation (5) is supposed continuous and subject to the Lipschitz condition in x. Then either Σ is empty or there exists in Σ a relaxed trajectory for which the difference at the endpoints of the coordinates x_0 of x assumes its minimum.

The Pontrjagin Maximum Principle: Integrated Version

For greater generality we suppress dependence on the chattering control v(t) by writing g(t, x) for g(t, x, v(t)). Consider a convex family G of such functions, a family such that every convex combination

$$\Sigma \alpha_i g_i, \tag{37}$$

of a finite number of members g_i, of G with constant coefficients $\alpha_i \geq 0$, where $\Sigma \alpha_i = 1$, is itself a member of G (in the chattering case the family G are functions g of the form g(t, x, v(t)) and convexity holds in a stronger sense in which the coefficients α_1 are allowed to be measurable functions of t instead of constants). In addition we require that every function g(t, x) in G is continuously differentiable in x for fixed t and measurable in t for fixed x, and also that each g and its partial derivative g_x are bounded functions of (t, x) or, more generally, bounded in absolute value by some integrable function of t only. These various requirements are to hold only in some bounded open set O of (t, x)-space. In the chattering case all these requirements are satisfied if we make the stipulation that g(t, x, u) is continuously differentiable.

We consider the family

$$H = yG \tag{38}$$

of Hamiltonian functions h(t, x, y) = yg(t, x), where y is a variable vector and each $g \in G$ gives rise to a corresponding $\lambda \in H$. we shall be concerned with points (t, x) that lie in a sufficiently fine neighborhood of the set described by a given fixed trajectory C of the form x(t), $t_1 \leq t \leq t_2$. In terming C a trajectory we imply that the function x(t) is, almost everywhere in its interval, a solution of the differential equation,

$$\dot{x} = g(t, x(t)), \tag{39}$$

for some fixed corresponding member $g \in G$; moreover, x(t) is to be absolutely continuous.

We term ordinary point of C a point at which (39) holds; in particular, we say that C has ordinary endpoints if the derivatives $\dot{x}(t_i)$ exist and have the values $g(t_i, x(t_i))$, i = 1, 2. We assume function x(t) continued outside its interval domain, when convenient, subject to the same differential equation and absolute continuity. In view of the uniqueness theorem any such extension is uniquely determined once we fix the member $g \in G$ and an initial condition of the type $x(t_0) = x_0$.

We write q for the ordered pair of endpoints of C and P for a small neighborhood of q. Thus, p lies in the space of such ordered pairs q, *i.e.*, in the Cartesian product of (t, x)-space with itself or, equivalently, O with itself. We denote by Q the subset of P consisting of ordered pairs (p, q) of endpoints of trajectories in O, sufficiently close to C. Any such trajectory has the form $\chi(t)$, $\tau_1 \leq \tau \leq \tau_2$, where $\chi(t)$ is absolutely continuous and satisfies, for almost all t in its interval of definition, a differential equation similar to (34) with g replaced by some member $\hat{g}$ of form g(t, x, v(t)).

In P suppose given a smooth manifold M with q as boundary point where M is represented by local coordinates as a smooth one-to-one map of a smooth Euclidean domain with its boundary. We take the interior of M and boundary of

M as corresponding images of the interior and boundary of this domain. We suppose the dimension of M to be ≥ 1 so that it does not reduce to a point. We suppose, further, that the boundary of M at q has a tangent subspace, which is itself the boundary of a tangent half-space (a tangent half-subspace by taking half-lines tangent to M at q). We suppose that a neighborhood of q in M has a continuous one-one map onto a neighborhood of q in this tangent half-subspace such that q corresponds to itself and that, if q + δp denotes the image of p in M, we have,

$$P = q + \delta p + o(p - q) \tag{40}$$

where o(p − q) is small compared with p − q as p → q. In particular, local coordinates can be thought of as coordinates on the tangent half-subspace. Moreover, a vector $\phi \neq 0$ in the underlying (2n + 2)-dimensional Euclidean space will be termed an inward normal of M at the point q if it is, first, orthogonal at q to the boundary of M, *i.e.*, to a hyperplane through q that contains the tangent subspace of this boundary and, second, directed towards the side of this hyperplane that contains the tangent half-subspace of M.

A trajectory C is an M-extremal if it contains no interior point of M; we term conjugate vector along C an absolutely continuous and nowhere vanishing vector-valued function y(t) with values in n-space defined on the same interval as x(t). If h is the Hamiltonian corresponding to the element $g \in G$ that enters into the differential equation (38) satisfied by x(t), we term corresponding momentum and denote by η(t) the (n+1)-dimensional vector derived from y(t) by the adjunction of the initial component

$$-h(t, x(t), y(t)) = \eta_0(t). \tag{41}$$

We term corresponding transversality vector for C the (2n+2)-dimensional vector,

$$(-\eta(t_1), \eta(t_2)). \tag{42}$$

The Integrated form of the Pontrjagin Maximum Principle

Let $g \in G$, h be the corresponding Hamiltonian function yg(t, x) and let C be an M-extremal of the form x(t), $t_1 \leq t \leq t_2$, satisfying, almost everywhere, the corresponding differential equation (21) which we now write $\dot{x} = \partial h/\partial y$. If C has ordinary endpoints or M consists of pairs with the same coordinates as q, then there exists a conjugate vector y(t) along C such that the pair (x(t), y(t)) satisfies the following three condition:

(a) The canonical Euler equations:

$$\dot{x} = \partial h/\partial y, \quad \dot{y} = -\partial h/\partial y. \tag{43}$$

(b) The Weierstrass condition: As function of $\lambda \in H$, the quantity,

$$\int_\Delta h(t, x(t), y(t))dt, \tag{44}$$

assumes its maximum when $\lambda = h$.

(c) The transversality condition:

Then the transversality vector (42) is an inward normal of M.

NEW MATHEMATICS

We shall consider here the new mathematics involved the development of GUT, namely, the new real number system, chaos and turbulence and generalized fractal, integral and derivatives but we focus on the new real number system and the generalized integral; the rest we shall summarize or refer to original sources. We have already introduced qualitative mathematics and qualitative modeling, the most important new mathematics involved in GUT. The new real number system provides qualitative models for important physical systems like the superstring and our universe. Therefore, it is a very important component of the new methodology.

Fractal, Generalized Fractal and Chaos

The generalized fractal of particular importance to GUT that we consider here is nested generalized fractal. (See [11, 48, 53, 87, 109] on geometrical fractal)

Definitions and Examples

Classical fractal is iterated affine transformation of a given generator, some geometrical figure in the case of a geometrical fractal; it is mainly quantitative [13, 24]. Affine transformation is a combination of contraction and translation the effect being to generate a sequence of self-similar figures, *i.e.*, each term except for the first is similar to the preceding term, at decreasing scale. We generalize this fractal construction to include as well, rotation, taking mirror image, sliding along a curve and replication all of which preserve similarity. The last characteristic is the most important for applications, especially, in biology. We have seen this replication in the form of splitting or branching of roots and branches of a tree and the veins of its leaf. This also happens in mitosis or cell-division where self-similarity is in terms of the replication of the genes in every offspring cell. The last two examples are physical fractals where the sequences involved are finite since visible physical systems are finite our visible universe being finite [25]. Moreover what is replicated need not be geometrical figure but general properties of the terms of the fractal sequence. We call this formation generalized fractal; only qualitative mathematics is capable of modeling generalized fractal, especially, when there is multiple replication or multifurcation. In the root of a tree every branch continues as nested fractal, *i.e.*, each term in the sequence is contained in or a part of the preceding term (in this case at decreasing scale). The characteristic here is replication more than self-similarity and decreasing scale because the terms may not be geometrical but some characteristics or processes as in the above examples of replication. Mitosis in a living cell is a special fractal that biologists use to describe its replication in the offspring cell, especially, its genetic content; here decreasing scale does not apply but self-similarity and replication do.

Nested fractal is nature's way of packing huge energy in a physical system or attaining maximum efficiency in a physical process, an expression of energy conservation and a universal configuration of nature that applies even to man-made structures such as machines, buildings and bridges due to its optimal properties. In biology, encoding of information in the brain is a fractal process [29].

The fractal structure of the roots of a tree allows it to absorb maximum nutrients from the ground and that of its branches and veins of its leaf allows optimal efficiency in the distribution of nutrients to the stomata for food production (photosynthesis), e.g., fruit, and distribution to where they are needed including fruits that humans can harvest.

Chaos is mixture of order none of which is identifiable. For example, in regions where there is much under-ocean volcanic activity the ocean surface heats up, throws the gas molecules above it into motion (kinetic energy) that pushes them apart and creates low pressure. Low pressure sucks gas molecules around it and the initial rush throws trillions of gas molecules into collision that makes it impossible to monitor or even predict the path of any single one. At the same time, every molecule is subject to the laws of nature. This is then a classic case of chaos but it is transitional since collision is energy dissipating. Energy conservation induces global order and turns it into coherence of order called turbulence in this case a hurricane.

The formation of tropical cyclone is an example of standard dynamic system. It starts on a calm summer day which is order; then chaos ensues as a transition to coherence of order called turbulence, in this case, a cyclone which is a vortex of gas molecules in the atmosphere. This transition from chaos to turbulence is due to energy conservation. Chaos is energy dissipating; in this example it is due to collision of gas atoms and molecules which is distortion of order; therefore, energy conservation induces its evolution into global order, the cyclone. Then the cyclone vanishes when nothing infuses energy on it, e.g., warm corridors on the ocean, and its energy dissipates into the atmosphere; when a cyclone passes through a warm corridor its power rises because heat or kinetic energy is fed to it. It was once suggested that an atom bomb be dropped into the eye of a typhoon to end it. That would be like pouring gasoline into brush fire. At the same time, when the eye hits and gets plugged in by a mountain its power declines because friction dissipates its energy.

It is impossible to model chaos computationally, only qualitative modeling can. Another example is fundamental chaos or dark matter one of the two fundamental states of matter that we shall consider later; fundamental in the sense that it is part of the cycle of matter: all matter comes from it and will ultimately return to it. However, fundamental chaos is not energy dissipating because the semi-and non-agitated superstrings do not interact among themselves; therefore, it is stable and has zero entropy, the only stable physical chaos. (The real line is mathematical chaos [24, 78]; so are infinitesimal zigzag and oscillation [24, 78]).

The Peano Space-Filling Curve

The problem here is to map the unit interval [0, 1] *onto a* unit square by continuous function. The usual tool is a theorem that says: the limit of a uniformly convergent sequence of continuous functions is continuous. But all constructions so far are difficult to visualize. We make the geometrical construction of this curve quite simple [27, 34] using the concept of nested fractal sequence.

Divide the unit square into 9 little squares by the lines,

$$x = 1/3, x = 2/3, y = 1/3, y = /2/3. \tag{45}$$

We label the blocks, B_1, B_2, B_3, B_4, B_5, B_6, B_7, B_8, B_9, starting at the bottom row from left to the right, then up to the middle, then left to the first block at the middle row, then up on this block to the top row, then right to the last block of the top row to the right hand corner of the square (see Fig. **4**); details are discussed below. We take as the first initial generator the segment $g_{1,1}(x)$: $y = x$, $0 \le x \le 1/3$; take the mirror image of (flip over) $g_{1,1}(x)$ with respect to the y-axis to obtain the second initial generator $g_{1,-1}(x)$: $y = -x$, $0 \le x \le 1/3$; take the mirror image of the second initial generator to obtain to obtain the third initial generator, $g_{-1,-1}(x)$: $y = x$, $0 \ge x \ge -1/3$ and take the mirror image of the third generator to obtain the fourth generator, $g_{-1,-1}(x)$: $y = x$, $0 \ge x \ge -1/3$. The four initial generators are:

$$g_{11}(x): y = x, 0 \le x \le 1/3; g_{-11}(x): y = -x, 0 \le x \le 1/3;$$

$$g_{-1-1}(x): y = x, 0 \ge x \ge -1/3; g_{1-1}(x): y = -x, 0 \ge x \ge -1/3. \tag{46}$$

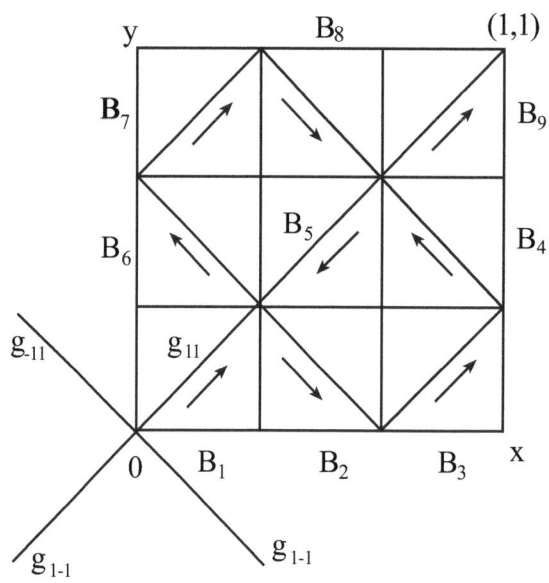

Figure 4: The first term $f_1(t)$ of the fractal sequence and its generators g_{11}, g_{-11}, g_{1-1} and g_{1-1} in the construction of the Peano space-filling curve obtained by translating them suitably to form the ordered polygonal line from the origin to the point (1, 1). Arrows indicate direction of the polygonal line.

Then the initial function $f_1(t)$ in the iteration process follows:

$$f_1(x): g_{11}(x), g_{1-1}(x), + (1/3.1/3), g_{11}(x) + (2/3),$$

$g_{-11}(x) + (1, 1/3)$, $g_{-1-1}(x) + (2/3, 2/3)$,

$g_{-11}(x) + (1/3, 1/3)$, $g_{11}(x) + (0, 2/3)$,

$g_{1-1}(x) + (1/3, 1)$, $g_{11}(x) + (2/3, 2/3)$, $0 \le x \le 1/3$.   **(47)**

The functions that comprise $f_1(x)$ are suitable translations of the generators in (43) to form a polygonal line through the diagonals of

blocks B_1, B_2,..., B_9, in that order with suitable orientation. Construction of $f_1(x)$:

(1)   The first segment of $f_1(x)$ is $g_{11}(x)$.

(2)   For the second segment in block **B_2**, use $g_{1-1}(t)$ by translating it to the right by 1/3 units and up by 1/3 units so that its initial point coincides with the terminal point of $g_{11}(x)$ and terminal point coincides with the lower right hand vertex of **B_2**.

(3)   For the second segment, simply translate $g_{11}(x)$ to the right by 2/3 units so that the initial point coincides with the terminal point of the previous segment and the terminal point with the upper right hand vertex of **B_3**.

(4)   Then we move up to **B_4**. The segment that will go here is suitable translation of $g_{-11}(x)$, etc. Then follow the same construction with suitable translations of generators through **B_5**, **B_6**, ..., **B_9** so that the end point of $f_1(x)$ coincides with the point (1, 1). The initial construction of $f_2(x)$ is shown in Fig. **5**.

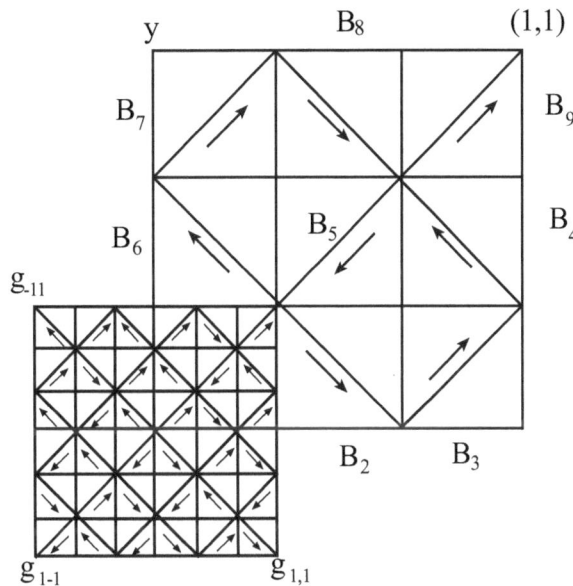

Figure 5: Contracting $f_1(t)$ by 1/3 at the origin yields one of the generators $g_{22}(t)$; the other generators are as follows: g_{-22}, mirror image of g_{22} about the y-axis, g_{-2-2}, mirror image of g_{-22} about the −x-axis and g_{2-2} that of g_{-2-2} about the −y-axis. Suitable translation of these generators forms a finer polygonal line from the origin to the point (1, 1), the second term $f_2(t)$ of the fractal sequence [71].

The second set of generators for the second term $f_2(x)$ of the fractal sequence is constructed as follows:

(1)   Contract f_1 through its projection cone with vertex at the origin and call its image g_{22}.

(2)   Flip g_{22} over the y-axis, *i.e.*, take its mirror image with respect to it, and call it g_{-22};

(3)   Flip g_{-22} over the −x-axis and call the image g_{-2-2};

(4)   Flip g_{-2-2} over the −y-axis and call the image g_{2-2}.

(Fig. **5** shows the construction of the second set of generators)

We construct $f_2(x)$ using the generators $g_{22}(x)$, $g_{-22}(x)$, $g_{-2-2}(x)$ and $g_{2-2}(x)$ as follows:

$f_2(x)$: $g_{22}(x)$, $g_{2-2}(x) + (1/3, 1/3)$, $g_{22}(x) + (2/3, 0)$,

$g_{-22}(x) + (1.1/3)$, $g_{22}(x) + (2/3, 0)$, $g_{-2-2}(x) + (2/3, 2/3)$,

$g_{-22}(x) + (1/3, 1/3)$, $g_{22}(x) + (0, 2/3)$,

$g_{2-2}(x) + (1/3, 1)$, $g_{22}(x) + (2/3, 2/3)$, $0 \le t \le 1/3$.   **(48)**

The next phase is, again, contraction of $f_2(x)$ by 1/3 to form the first of the four generators of f_3 denoted by $g_{33}(x)$ and taking suitable mirror images to complete its four generators. Then we iterate this phase of the construction to generate the sequence of functions, $f_1(x)$, $f_2(x)$,..., $f_n(x)$, with the following properties:

(1)   Each $f_n(x)$, n = 1, 2,..., is made up of polygonal lines as suitable translations of the generators of $f_n(x)$;

(2)   Self-similarity is obvious from the construction since the orientation of $f_n(x)$, for each n, is preserved;

(3)   Each $f_n(x)$ in the sequence is continuous;

(4)   The sequence $f_n(x)$, n = 1, 2,..., is uniformly convergent;

(5)   Therefore, the sequence $f_n(x)$, n = 1, 2,..., converges to a continuous function f(x) whose graph clearly fills up the unit square.

Filling up the Unit Cube

The above construction can be extended to fill up the unit cube by the continuous mapping of the unit interval as follows:

(1)   Consider the unit cube with vertices A(0, 0, 0), B(1, 0, 0), C(0, 1, 0), D(0, 0, 1).

(2)   Subdivide the cube by the planes,

$x = 1/3$, $x = 2/3$, $y = 1/3$, $y = 2/3$, $z = 1/3$, $z = 2/3$,   **(49)**

into 27 little cubes.

(3)   As in the construction of the Peano space-filling curve, connect the diagonals of the little cubes from A(0, 0, 0) suitably in the right order to form a polygonal line through the little cubes once and ending up at so that the entire cubes is covered stretch and deform the unit interval AB to a polygonal line and map each segment in suitable order into the diagonals of at the origin joining the vertices B(0, 0, 0) and B(1/3, 1/3, 1/3),..., B(2/3, 2/3, 2/3) and B(1, 1, 1), where the last segment is mapped into the diagonal of the little cube opposite the first little cube at the origin and the terminal point at the opposite corner of the cube at its diagonal. Call this polygonal line f_1.

(4)   Contract f_1 to 1/3 into the first cube (by pushing it through its projection cone with vertex at the origin. This contracted polygonal line becomes the first generator g_{111} of f_2; take the mirror image of g_{111} about the y-axis; the image is the second generator g_{1-11} of f_2; take the mirror image of g_{1-11} about $-x$-axis to obtain the third generator g_{1-1-1}; take the mirror image of g_{1-1-1} about the $-y$-axis to obtain the fourth generator of f_2.

(5)   Suitably translate these generators as in the construction of f_1 to form the appropriate finer polygonal line from the origin to the point (1, 1) of the unit square made up of the segments of f_1 through the contracted cubes with the terminal point at the vertex of the original cube opposite the origin.

(6)   Continue this iteration procedure to obtain a uniformly convergent sequence of continuous functions f_1, f_2, ... whose point-wise limit is a continuous function. We shall call this the Peano cube space-filling curve.

This construction is constructivist, intuitive and the simplest so far. Earlier constructions use this theorem: the limit of a uniformly convergent sequence of continuous functions is continuous. This construction can be extended to the n-hypercube where n is odd.

The Infinitesimal Zigzag as Limit Set of Fractal

Consider without loss of generality the equilateral triangle ADB of Fig. **1** with one vertex at the origin A and with base AB. We denote by $g_{11}(t)$ side AD which we take as the initial generator. We take the mirror of $g_{11}(t)$ with respect to the y-axis and denote it by $g_{-1,1}(t)$ as the second generator. The third generator is the mirror image of the second with respect to the negative x-axis denoted by $g_{-1-1}(t)$ and the fourth generator is the mirror image of the third with respective to the negative y-axis denoted by $g_{1-1}(t)$. Their parametric equations are given by:

$$g_{11}(t): (x, y) - (t, t),\ g_{1-1}(t) : (x, y) = (y, -t),\ 0 \le t \le 1/2. \tag{50}$$

To construct function $f_1(t)$, we take $g_{11}(t)$, $0 \le t \le \frac{1}{2}$, as part of $f_1(t)$ and combine it with the translation of $g_{1-1}(t)$ given by $(x, y) = g_{1-1}(t) + (1/2, 1/2)$, $0 \le t \le 1/2$. Its graph is the polygonal line formed by segments AD and DB.

To construct $f_2(t)$ in the iteration process, we contract $g_{11}(t)$ by $\frac{1}{2}$ and denote this by $g_{22}(t)$. This is one of the generators. Rotate $g_{22}(t)$ by $-\pi/2$ to form another generator $g_{2-2}(t)$. The generators are given by $g_{22}(t): (x, y) = (t, t)$, $g_{2-2}(t) = (t, -t)$, $0 \le t \le 1/4$.

The function $f_2(t)$ is given by the following system of equations,

$$f_2(t): (x, y) = g_{22}(t),\ (x, y) = g_{2-2}(t) + (1/3, 1/4),$$

$$(x, y) = g_{22}(t) + (1/2, 0),\ (x, y) = g_{2-2}(t) + (3/4, 1/4),$$

$$0 \le t \le 1/4, \tag{51}$$

and is represented by the polygonal line APQRB with x replaced by t and y replaced by x.

We iterate this construction to generate a sequence of functions $f_n(t)$ represented by the curve C_n, n = 1, 2,..., with the following properties:

(1)　The sequence $f_n(t)$, n = 1, 2,..., is uniformly convergent and each f_n is continuous,

(2)　For each n, the length $|C_n|$ of C_n satisfies

$$|C_n| = |C_1| = \sqrt{2}, \tag{52}$$

(3) The sequence f_n converges to a continuous function $f_n(t)$ represented by the curve C_0: $y_0(t) = 0$ which coincides point-wise with the ordinary constant curve C: $x(t) = 0$, $0 \le t \le 1$, distinct from it since

$$|C_0| = \lim |C_n| = \lim |\sqrt{2}| = \sqrt{2} \ne |C|. \tag{53}$$

This curve C_0 is the infinitesimal zigzag.

One property of the superstring is that, left alone, it shrinks steadily. We model this behavior mathematically using the superposition of a sinusoidal curve over a polygonal line as shown in [24, 34] (the sinusoidal curve is the projection of the helix on the plane through its axis). Given any real number r > 1, there exists an isosceles triangle ADB with sides AD and DB having total length $|AD| + |DB| = r$. Following the above construction, we form a sequence of polygonal lines whose limit in the sup norm is $|AB|$ but whose length is r. Thus, AB is a coincidence of countably many curves distinct from AB and from each other. The ordinary segment AB is the visible element of the countably infinite space of generalized fractals. The rest is dark matter.

The preceding construction can be generalized to apply to any triangle, ADB, where the slopes of AD and DB are m_1 and m_2, respectively, and $-\infty < m_1 < \infty$, $-\infty < m_2 < \infty$. Then we can pack countably infinite infinitesimal zigzags into AB whose lengths vary along all values in the interval $|AB|, \infty)$ each of which has set-valued derivative (m_1, m_2) at each point on AB.

A More General Geometrical Fractal

We first illustrate a convenient way of contracting a compacts set.

Given compact set B in $\mathbf{R}^n$ (subspace of the n-Cartesian product of the real line) and any real number s, $0 \le s \le 1$, then the set,

$$sB = \{sb \mid b \in B\}, \tag{54}$$

is similar to B in the sense that for any point $b \in B$ with components bk, $k = 1, 2,..., n$, $sbk/bk = s$. *i.e.*, ratios of components are preserved. For a plane set B (Fig. **6**) the set sB is a contraction of B along the projection cone B* towards its vertex at the origin where remains in the line segment joining it to the origin during contraction. In fact, B* can be expressed as

$$B^* = \cup\{tB\}, 0 \le t \le 1, \tag{55}$$

obtained by taking the union of B with its projection cone whose vertex is the origin. Note that since for any real number s, $0 < s < 1$, $\lim s^m = 0$, the compact set B in $\mathbf{R}^n$ can be shrunk towards a point at the origin by iterated construction,

$$sB, s^2B, ..., s^mB,.... \tag{56}$$

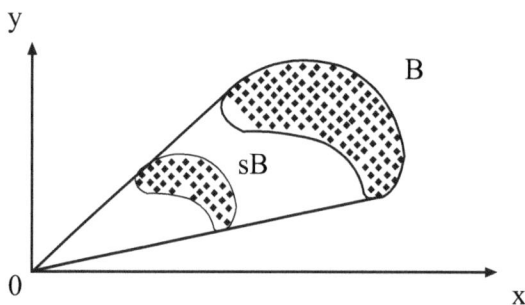

Figure 6: Plane geometrical figure B contracted by s by pushing it suitably through its projection cone towards its vertex at the origin.

Suppose B has diameter of length δ, where

$$\delta = \sup\{d(p,q) \mid p,q \in B\} \tag{57}$$

and $d(p, q)$ is the Euclidean distance between p and q. Let $s = 1/3$. We rotate and translate B/3 suitably so that its diameter coincides with the x-axis and its left extreme point is at the origin. We translate B/3 along the x-axis twice one at a time so that their images join end to end at extreme points shown in Fig. **7a**. Consider B/3 at the origin and its two images and denote them by B_{11}, B_{12}, B_{13}, respectively; denote their union by

$$B_1 = B_{11} \cup B_{12} \cup B_{12}. \tag{58}$$

B_1 is the generator and first term of the nested fractal sequence we are constructing. We contract B_1, again, through its projection cone by 1/3. The image consists of contraction of B/3 to $B/3^2$ at the origin and similar translation of the latter twice one at a time so that they join end to end as in the previous construction shown in Fig. **7b**. We denote the components of the contraction of B_1 by B_{21}, B_{22}, B_{33}, respectively, where each is the contraction of B/3 to $B/3^2$. Note that the image of B_1 lies inside B_{11} and the extreme right endpoint of B_{23} coincides with the extreme right endpoint of B_{13}. We translate B_{11} (which contains B_1) twice one at a time along the x-axis so that their images are end to end with B_{11} and denote them by B_{21}, B_{22}, B_{23} each component of which is equal to the contraction of B/3 to $B/3^2$. Then the second term of our nested geometrical fractal sequence is given by

$$B_2 = B_{21} \cup B_{22} \cup B_{23}. \tag{59}$$

We iterate the construction and form the nested fractal sequence,

$$B_n, n = 2, 3, \ldots. \tag{60}$$

and take k successive translations of $g_{11} = \delta B$:

$$G_{11} = g_{11}$$

$$G_{12} = g_{11} + (\delta/k, 0, \ldots, 0)$$

$$G_{13} = g_{11} + (2\delta/k, 0, \ldots, 0)$$

$$G_{1k} = g_{11} = ((k-1)\delta/k, 0, \ldots, 0). \tag{61}$$

Take the union of the G_{1m}, m =1, 2, ..., k – 1, to obtain,

$$G_2 = \cup G_{1m}, m = 1, 2, \ldots, k - 1. \tag{62}$$

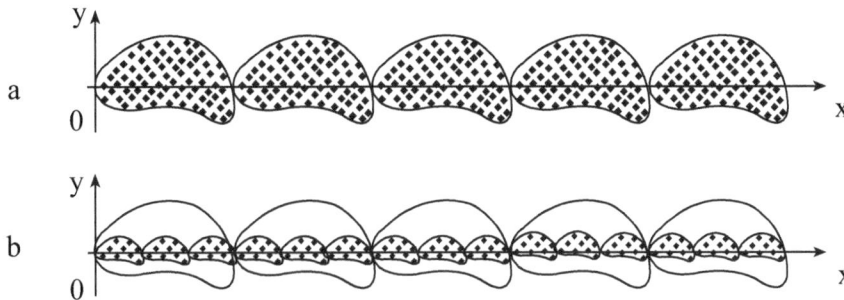

shown in Fig. **7** for a three-component generator.

We iterate this construction to obtain a sequence $G_1, G_2, \ldots, G_m$, forming a nested geometrical fractal that shrink to the x-axis joining the origin and the point $(\delta, 0, \ldots, 0)$. Again, obvious generalization can be done by allowing linear combination of the different contractions of the generators at each stage in the affine transformation.

We note that the set limit, point-wise or in the sup norm, of a nested geometrical fractal set is chaos.

This method can be further generalized by allowing linear combination of different contractions of the generator at each stage in the iteration process.

This fractal mathematics, particularly, infinitesimal oscillation is the key to an understanding of physical singularities such as black hole and the tremendous but dark or latent energy in the nucleus of an atom. As we shrink an oscillation to a point with its length preserved, its energy hv, where h is the Planck's constant and v its frequency, rises without bounds. However, it is dark (undetectable) with respect to our present means of observation such as light due to difference in orders of magnitude between their frequencies (the same principle that applies to non-resonance of radio or TV reception).

Theorem 3 is really a prescription for pushing matter (of which oscillation is its universal motion and fractal its universal configuration) into the hidden or dark region of matter. It is quite well known in physics today that 95% of matter in our universe is dark [8, 42]. Dark matter consists of non-agitated superstrings, visible matter of agitated superstrings; a primum is agitated superstring [43].

Note that the proofs of the above theorem are geometrical and much simpler and suitable for animation. In fact, fractal construction of the Peano space-filling curve was animated in the presentation of [44]. This construction is an extension of the method used in the Peano space-filling curve as limit of nested fractal sequence.

Fractal is everywhere in nature because it is one of the expressions of energy conservation identified in the law of nature called Energy Conservation Equivalence. It is nature's way of packing huge energy in a physical system or carrying out a process most efficiently.

The New Real Number System

We now optimize the applications of qualitative mathematics to rectify the weakness of the real number system and build the new real number system as the new foundation of mathematics that retains all the interesting and desirable properties of the real numbers.

Our Strategy

Our strategy is not simply to build the contradiction-free mathematical space called the new real number system **R*** but also to meet the needs of natural science and practical affairs while retaining the valid interesting and useful properties of the real number system. Then the new real number system must contain mathematics that has worldwide applications. In particular, it must provide both continuous and discrete mathematics including the decimals whose physical model, the metric system, has worldwide applications that other systems of measures are converting to it. Any other useful mathematics that may arise we shall consider a bonus. Concretely, the new real number system must be a continuum since physical space is which pervades everything and cannot be split into disjoint nonempty subsets. The key is to choose the right consistent axioms upon which to build it. These are the parameters for our construction.

The Terminating Decimals

We first build our base space, the terminating decimals **R** under these axioms:

Axiom 1. 0 and 1 are elements of **R**.

Axioms 2 and 3. The addition and multiplication tables.

Axioms 2 and 3 initially well-define 0 and 1 then the digits or basic integers 0, 1, 2, 3, 4, 5, 6, 7, 8 and 9 and the terminating decimals.

They are the elements of **R** as a mathematical space. The nonterminating decimals belong to the extension of **R** called the new real number system denoted by **R***. The elements 0 and 1 are called the additive and multiplicative identities of **R**, respectively. Initially, they are ill-defined until axioms 2 and 3 well-define them as well as their properties and relationship with the other integers and the terminating decimals.

We first define the digits or basic integers beyond 0 and 1:

$$1 + 1 = 2; 2 + 1 = 3; ..., 8 + 1 = 9. \qquad (63)$$

We omit the statements of the addition and multiplication tables which are familiar to everyone since primary school. Then we define the rest of the integers as base 10 place-value numerals:

$$a_n a_{n-1} ... a_1 = a_n 10^n + a_{n-1} 10^{n-1} + ... + a_1, \qquad (64)$$

where the a_ns are basic integers. The metric system models the system of decimals.

Now, we extend the integers to include the additive and multiplicative inverses $-x$ and, if x is not 0, $1/x$ (reciprocal of x), respectively. Note that the reciprocal of an integer exists only if it has no prime factor other than 2 or 5. We also extend the operations $+$ and $\times$ by re-stating associativity, commutativiy, distributativy, etc., and introduce something else that is new: the rules of sign that we take as part of the axioms of this extension (we need not write them as they are familiar). Then we define subtraction as a new operation: the difference between x and y or y subtracted from x. Then we define another new operation: division of an integer x by a nonzero integer y, or quotient, denoted by x/y and defined by:

$$x/y = x(1/y), \tag{65}$$

provided y is neither 0 nor a pime other than 2 and 5. We similarly extend distributivity of multiplication relative to addition and include them as axioms of the extension. We consider subtraction the inverse operation of addition and division that of multiplication as examples of duality that we shall consider in detail below. Formally, we define subtraction of y from x by the equation: $x - y = x + (-y)$ and division by: $x/y = x(1/y)$. We define a terminating decimal as follows:

$$a_n a_{n-1}\ldots a_1.b_k b_{k-1}\ldots b_1 = a_n 10^n + a_{n-1}10^{n-1} +\ldots + a_1 + b_1/10$$

$$+ b_2/10^2 +\ldots + b_k/10^k = a_n 10^n + a_{n-1}10^{n-1} +\ldots + a_1$$

$$+ b_1(0.1) + b_2(0.1)^2 + \ldots + b_k(0.1)^k. \tag{66}$$

where $a_n a_{n-1}\ldots a_1$ is the integral part, $b_1 b_2\ldots b_k$ the decimal part and $0.1 = 1/10$. The terminating decimals are well-defined since the reciprocal of 10 has only the factors 2 and 5. If x and y are relatively prime integers, $y \neq 0$, then the quotient x/y of x by y exists only if y has no prime factor other than 2 or 5. Such quotient is called rational.

The Nonterminating Decimals

We define the nonterminating decimals for the first time without contradiction and with contained ambiguity, *i.e.*, approximable by certainty. We build them on what we know: the terminating decimals, our point of reference for all its extensions.

A sequence of terminating decimals of the form,

$$N.a_1, N.a_1 a_2, \ldots, N.a_1 a_2\ldots a_n, \ldots. \tag{67}$$

where N is integer and a_n is called standard generating or g-sequence. Its nth g-term, $N.a_1 a_2\ldots a_n$, defines and approximates its g-limit, the nonterminating decimal,

$$N.a_1 a_2\ldots a_n\ldots, \tag{68}$$

at margin of error 10^{-n} provided each nth g-term is computable, *i.e.*, there is some algorithm or rule for computing the nth digit from the digits. For example, the nth digit can be the last digit of the sum of the squares of its preceding two digits. The digits of π can be computed from its infinite series expansion. A decimal is normal if every digit is chosen at random the digits [16]. The g-limit of (67) is the nonterminating decimal (68) provided the nth digits are not all 0 beyond a certain value of n; otherwise, it is terminating. As in standard analysis where a sequence converges, *i.e.*, tends to a number, in the standard norm, a standard g-sequence, converges to its g-limit in the g-norm where the g-norm of a decimal is itself. Note that a decimal consists of the integral part, the integer to the left of the decimal point, and the decimal part, the sequence of digits to the right of the decimal point which may be terminating or nonterminating. Then we alternatively define an integer as the integral part of a decimal.

We recall that in the real number system a rational is defined as nonterminating periodic, *i.e.*, the digits are periodic after a certain digit, and a real number is irrational if it is nonterminating and nonperiodic. Each of these concepts is ambiguous for it is impossible to verify if the digits are periodic or not. Thus, the concept *irrational* is ambiguous which we discard so that the decimals belong to two mutually exclusive classes, terminating or rational and nonterminating.

We define the nth distance d_n between two decimals a, b as the numerical value of the difference between their nth g-terms, a_n, b_n, *i.e.*, $d_n = |a_n - b_n|$ and their g-distance is the g-limit of d_n. We denote by **R*** the g-closure of **R**, *i.e.*, its closure in the g-norm.

A terminating decimal is degenerate nonterminating decimal, *i.e.*, the digits are all 0 beyond the nth decimal digit for some n. The nth g-term of a nonterminating decimal repeats every preceding digit at the same order so that if finite terms are deleted the nth g-term and g-limit are unaltered and the remaining terms comprise its g-sequence. Thus, a nonterminating decimal may have many g-sequences and we consider them equivalent for having the same g-limit.

Since addition and multiplication and their inverse operations subtraction and division are defined only on terminating decimals computing nonterminating decimals is done by approximating each term or factor by its nth g-term (called n-truncation) which is a terminating decimal and using their approximation to find the nth g-term of the result of addition or multiplication and its inverse operation as its approximation at the same margin of error. This is standard computation, *i.e.*, approximation by decimal segment at the nth digit. Thus, with our premises we have retained standard computation but avoided the contradictions and paradoxes of the real numbers. We have also avoided vacuous statement, e.g., vacuous approximation, because nonterminating decimals are g-limits of g-sequences which belong to **R***. Moreover, we have contained the inherent ambiguity of nonterminating decimals by approximating them by their nth g-terms which are not ambiguous being terminating decimals. In fact, the ambiguity of **R*** has been contained altogether by its construction on the additive and multiplicative identities 0 and 1.

As we raise n, the tail digits of the nth g-term of any decimal recedes to the right indefinitely, *i.e.*, it becomes steadily smaller until it is unidentifiable from the tail digits of the rest of the decimals. While it tends to 0 in the standard norm it never reaches 0 in the g-norm since the tail digits are never all equal to 0; it is also not a decimal since the digits are not fixed. Since none of the tail digits of a decimal is distinguishable from the rest the set of the tail digits of this set cannot be split into two distinct subsets which makes it a continuum in the algebraic sense.

In iterated computation to get closer and closer approximation of a decimal, e.g., calculating $f(n) = (2n^4+1)/3n^4$, n = 1, 2, ..., the tail digits may vary but recede to the right indefinitely and become steadily smaller leaving fixed digits behind that define a decimal. We approximate the result by taking its initial segment, the nth g-term, to desired margin of error by choosing n suitably.

The Dark Number d*

Consider the sequence of decimals,

$$(\delta)^n a_1 a_2 ... a_k, \ n = 1, 2, ..., \tag{69}$$

where δ is any of the decimals, 0.1, 0.2, 0.3, ..., 0.9, a_1, ..., a_k, basic integers (not all 0 simultaneously). We call the nonstandard sequence (68) d-sequence and its nth term nth d-term. For fixed combination of δ and the a_js, j = 1, ..., k, in (68) the nth d-term is a terminating decimal and as n increases indefinitely it traces the tail digits of some nonterminating decimal and becomes smaller and smaller until it is indistinguishable from the tail digits of the other decimals. As n → ∞ the nth d-term recedes to the right and tends to some number d, its d-limit in the d-norm, which is never 0 (since the a_js are not simultaneously 0 and each d-term is not 0). It is called dark number d which is indistinguishable from the rest of the d-limits of (68) for all other choices of δ and a_js. Therefore, the set of all dark numbers for all choices of δ and a_js is a countable continuum (since any set of sequences is countable) denoted by d*. Thus, d* is set-valued and a continuum (negation of discrete) of dark numbers; the decimals are joined by the continuum d* at their tails. The dark number d* is a continuum in the algebraic sense since no notion of disjoint open set is involved. Note that while the nth d-term of (69) becomes smaller and smaller with indefinitely increasing n it is greater than 0 no matter how large n is so that if x is a decimal, 0 < d < x. If an equation or function is satisfied by every dark number d we may substitute d* for d in it so that we can write 0 < d* < x in the above inequality.

At the same time, since the tail digits of all the nonterminating decimals form a countable combination of the basic digits 0, 1, ..., 9 they are countably infinite, *i.e.*, in one-one correspondence with the integers but their d-limits, being a continuum, have no cardinality (which applies only to discrete set). Any set whose elements can be labeled by integers or there is a scheme for establishing one-one correspondence between them and the integers is countably infinite. It follows that the countable union of countable set is countable.

Observation

Cantor's diagonal method proves neither the existence of nondenumerable set nor that of a continuum; it proves only the existence of countably infinite set, *i.e.*, the off-diagonal elements consisting of countable union of countably infinite sets. The off diagonal elements are not even well-defined because we know nothing about their digits (a decimal is determined by its digits). Therefore, we have the following:

Corollary

(1) Nondenumerable set does not exist; (2) Only discrete set has cardinality; a continuum has none.

Corollary (1) follows from the fact that a well defined set can be constructed only from at most countable union of finite set. Thus, the continuum hypothesis of set theory collapses. In view of the requirements of a mathematical space that it must be well defined by consistent set of axioms, it is not necessary to develop set theory as a kind of universal language for mathematics since its axioms are not valid in any mathematical space anyway unless there is a set of consistent axioms that well defines it in which case it becomes a mathematical space.

Like a nonterminating decimal, an element of d^* is unaltered if finite d-terms are altered or deleted from its d-sequence. When $\delta = 1$ and $a_1 a_2 ... a_k = 1$ (69) is called the basic or principal d-sequence of d^*, its d-limit the basic element of d^*; basic because all its d-sequences can be derived from it. The principal d-sequence of d^* is,

$$(0.1)^n, \; n = 1, 2, ..., \tag{70}$$

obtained from the iterated difference,

$$N - (N - 1).99... = 1 - 0.99... = 0, \text{ excess remainder of } 0.1;$$

$$0.1 - 0.09 = 0, \text{ excess remainder of } 0.01;$$

$$0.01 - 0.009 = 0, \text{ excess remainder of } 0.001; \tag{71}$$

Taking the nonstandard g-limits of the extreme left side of (70) and recalling that the g-limit of a decimal is itself and denoting by d_p the d-limit of the principal d-sequence on the rightmost side we have,

$$N - (N - 1).99... = 1 - 0.99... = d_p. \tag{72}$$

Since all the elements of d^* share its properties then whenever we have a statement "an element d of d^* has property P" we may write "d^* has property P", meaning, this statement is true of every element of d^*. This applies to any equation involving an element of d^*. Therefore, we have,

$$d^* = N - (N - 1).99... = 1 - 0.99.... \tag{73}$$

Like a decimal, we define the d-norm of d^* as d^* and $d^* > 0$.

Theorem

The d-limits of the indefinitely receding to the right nth d-terms of d^* is a continuum that coincides with the g-limits of the tail digits of the nonterminating decimals traced by those nth d-terms as the a_ks vary along the basic digits.

If x is nonzero decimal, terminating or nonterminating, there is no difference between $(0.1)^n$ and $x(0.1)^n$ as they become indistinguishably small, *i.e.*, as n increases indefinitely. This is analogous to the sandwich theorem of calculus that says, $\lim(x/\sin x) = 1$, as $x \to 0$; in the proof, it uses the fact that $\sin x < x < \tan x$ or $1 < x/\sin x < \sec x$ where both extremes tend to 1 so that the middle term tends to 1 also. In our case, if $0 < x < 1$, $0 < x(0.1)^n < (0.1)^n$ and both extremes tend to 0 so must the middle term and they become indistinguishably small as n increases indefinitely. If $x > 1$, we simply reverse the inequality and get the same conclusion. Therefore, we may write, $xd_p = d_p$ (where d_p is the principal element of d^*) and since the elements of d^* share this property we may write $xd^* = d^*$, meaning, that $xd = d$ for every element d of d^*. We consider d^* the equivalence class of its elements. In the case of $x + (0.1)^n$ and x, we look at the nth g-terms of each and, as n increases indefinitely, $x + (0.1)^n$ and x become indistinguishable. Now, since $(0, 1)^n > ((0.1)^m)^n > 0$ and the extreme terms both tend to 0 as n increases indefinitely, so must the middle term tend to 0 so that they become indistinguishably small (the reason d^* is called dark for being indistinguishable from 0 yet greater than 0). We summarize our discussion as follows: if x is not a decimal integer (a decimal integer has the form, $x = N.99...$, $N = 0, 1, ...$) then,

$x + d^* = x$; otherwise, if $x = N.99...$,

$x + d^* = N+1$, $x - d^* = x$; if $x \neq 0$, $xd^* = d^*$; $(d^*)^n = d^*$, $n = 1, 2, ..., N = 0, 1, ...$; $1 - d^* = 0.99...$;

$N - (N - 1).99...$; $1 - 0.99... = d^*$, $N = 1, 2, ...$ (74)

It follows that the g-closure of **R**, *i.e.*, its closure in the g-norm, is **R*** which includes the additive and multiplicative inverses and d^*. We also include in **R*** the upper bounds of the divergent sequences of terminating decimals and integers (a sequence is divergent if the nth terms are unbounded as n increases indefinitely, e.g., the sequence 9, 99, ...) called unbounded number u^* which is countably infinite since the set of sequences is. We follow the same convention for u^*: whenever we have a statement "u has property P for every element u of u^*" we can simply say "u^* has property P). Then u^* satisfies these dual properties: for all x,

$x + u^* = u^*$; for $x \neq 0$, $xu^* = u^*$.   (75)

Neither d^* nor u^* is a decimal and their properties are solely determined by their sequences. Then d^* and u^* have the following dual or reciprocal properties and relationship:

$0d^* = 0$, $0/d^* = 0$, $0u^* = 0$, $0/u^* = 0$, $1/d^* = u^*$, $1/u^* = d^*$.   (76)

Numbers like $u^* - u^*$, d^*/d^* and u^*/u^* are still indeterminate but indeterminacy is avoided by computation with the g- or d-terms.

It is clear that d^* and u^* are the counterparts of the infinitesimal and infinity of calculus; the only difference is that both d^* and u^* are well defined.

The decimals are linearly ordered by the lexicographic ordering "<" defined as follows: two elements of **R** are equal if corresponding digits are equal. Let

$N.a_1a_2..., M.b_1b_2... \in \mathbf{R}$.   (77)

Then,

$N.a_1a_2... < M.b_1b_2$ if $N < M$ or if N

$= M$, $a_1 < b_1$; if $a_1 = b_1$, $a < b_2$; ...,   (78)

and, if x is any decimal we have,

$0 < d^* < x < u^*$.   (79)

The trichotomy axiom follows from lexicographic ordering. This is the natural ordering mathematicians sought among the real numbers but it does not exist there because it contradicts the trichotomy axiom.

Mathematical Duals

Mathematical systems are better understood by bringing in the notion of dual systems as it introduces symmetry that may be useful. We look at divergent sequences, *i.e.*, sequences whose terms become bigger and bigger that they become indistinguishable from each other, as dual of convergent sequences. In this sense they also form a continuum. We denote their upper bounds by u* which satisfies (75), (76) and (79). Then we look at d* as the dual of u* and **R*** that of the system of additive and multiplicative inverses (except that it has holes, namely, the nonexistent multiplicative inverses of some primes). Thus, **R*** is a semi-field, the nonzero integers forming a semi-ring since some of them have no multiplicative inverses. Like d*, u* cannot be separated from the decimals, *i.e.*, there is no boundary between either of them and the decimals and between finite and infinite. Thus, we cannot separate d* from a decimal and there is no boundary to cross between finite and infinite so that beyond a certain finite decimal everything else is infinite. The latter is what is meant by the expression u* + x = u* for any decimal x. Duality is also seen in this case: let $\lambda > 1$ be terminating decimal then the sequence λ^n, n = 1, 2, ..., diverges to u* but $(1/\lambda)^n$, n = 1, 2, ... converges, d-lim $(1/\lambda)^n = d^*$.

Isomorphism Between the Integers and Decimal Integers

To find out more about the structure of **R*** we show the isomorphism between the integers and decimal integers, *i.e.*, integers of the form,

$$N.99..., N = 0, 1, ..., \tag{80}$$

but before doing so we first note that 1 + 0.99... is not defined in **R** since 0.99... is nonterminating but we can write 0.99... = 1 – d* so that 1 + 0.99... = 1 + 1 – d* = 2 – d* =1.99...; we now define 1 + 0.99... = 1.99... or, in general, N – d* = (N–1).99... Twin integers are pairs (N, (N–1).99...), N = 1, 2,...; the first and second components are isomorphic, the second called decimal integers.

Let f be the mapping N → (N – 1).99... and extend it to the mapping d* → 0 even if d* is not a decimal; then we show that f is an isomorphism between the integers and decimal integers.

(a) f(N+M) = (N+M–1).99... = N + M – 1 + 0.99...

= N – 1 + M – 1 + 1.99...

= N – 1 + 0.99... + M – 1 + 0.99...

$$= (N–1).99... + (M–1).99... = f(N) + f(M). \tag{81}$$

Thus, addition of decimal integers is the same as addition of integers. Next, we show that multiplication is also an isomorphism.

(b) f(NM) = (NM–1).999... = NM – 1 + 0.99...

= NM – N – M + 1 + N + –1 + M + –1 + 0.99...

= NM – N – M + 1 + (N–1).99... + (M–1).99...

+ (–1)(0.99...

= NM – N – M + 1 + N(0.99... + (–1)(0.99...

+ M(0.99... + (–1)(0.99...) + 0.99...

= (N – 1)(M – 1) + (N–1)(0.99...)

+ (M – 1)(0.99... + (0.99...)^2

$$= ((N-1) + 0.99... \ M-1) + 0.99...$$

$$= (((N-1).99...)((M-1).99...) = (f(N))(f(M)). \tag{82}$$

We have now established the isomorphism between the integers and the decimal integers with respect to both operations so that both subspaces of $\mathbf{R}^*$ are integers in the sense of [7]. We include in this isomorphism the map $d^* \to 0$, so that its kernel is the set $\{d^*, 1)$ from which follows equations (83):

$$(d^*)^n = d^* \text{ and } (0.99...)^n = 0.99..., n = 1, 2, ... \tag{83}$$

(The second equation can be proved also by mathematical induction)

For the curious reader we exhibit other properties of 0.99... Let K be an integer, M.99... and N.99... decimal integers. Then

(a) $K + M.99... = (K+M).99...,$

(b) $K(M.99...) = K(M + 0.99...) = KM + K(0.99...),$

(c) $M.99... + N.99... = M + N + 0.99... + 0.99...). \tag{84}$

To verify that $2(0.999...) = 1.99...$, we note that $(1.99...)/2 = 0.99...$

(d) $(M.99...)(N.99...) = (M + 0.99...)(N + 0.99...)$

$= MN + M(0.99... + N(0.99...) + (0.99...)^2$

$= MN + (M-1).999... + (N-1).99... + 0.99...$

$= MN + (M + N-2).99... + 0.99...$

$= MN + (M + N-.1).99... = (MN+M+N-1).99...,$

(e) $0.99... + 0.99... = 2(0.99...) = 1.99.... \tag{85a}$

Adjacent Decimals and Recurring 9s

Two decimals are adjacent if they differ by d^*. Predecessor-successor pairs and twin integers are adjacent. In particular, 74.5700... and 74.5699... are adjacent.

Since the decimals have the form $N.a_1a_2...a_n, ...,$ $N = 0, 1, 2, ...,$ the digits are identifiable and, in fact, countably infinite and they are linearly ordered by lexicographic ordering. Therefore, they are discrete or digital and the adjacent pairs are also countably infinite. However, since their tail digits form a continuum, $\mathbf{R}^*$ is a continuum with the decimals its countably infinite discrete subspace.

A decimal is called recurring 9 if its tail decimal digits are all equal to 9. For example, 4.3299... and 299.99... are recurring 9s; so are the decimal integers. (In an isomorphism between two algebraic systems, their operations are interchangeable, *i.e.*, they have the same algebraic structure but differ only in notation).

The recurring 9s have interesting properties. For instance, the difference between the integer N and the recurring 9, $(N - 1).99...,$ is d^*; such pairs are called adjacent because there is no decimal between them and they differ by d^*. In the lexicographic ordering the smaller of the pair of adjacent decimals is the predecessor and the larger the successor. The average between them is the predecessor. Thus, the average between 1 and 0.99... is 0.99... since $(1.99...)/2 = 0.99...;$ this is true of any recurring 9, say, 34.5799... whose successor is 34.5800... Conversely, the g-limit of the iterated or successive averages between a fixed decimal and another decimal of the same integral part is the predecessor of the former.

Since adjacent decimals differ by d* and there is no decimal between them, *i.e.*, we cannot split d* into nonempty disjoint sets, we have another proof that d* is a continuum (in the algebraic sense). Then we have another proof that **R*** is a continuum (also in the algebraic sense).

The counterexample to the trichotomy axiom shows that an irrational number cannot be expressed as limit of sequence of rationals since the closest it can get to it is some rational interval which still contains some rationals whose relationship to it is unknown, another expression of the fact that the concept *irrational* is ill-defined.

The g-sequence of a nonterminating decimal gets directly to its g-limit, digit by digit. Moreover, a nonterminating decimal is an infinite series of its digits:

$$N.a_1a_2...a_n... = N +.a_1 +.0a_2 + ... +.00...0a_n +...; 0.99...$$

$$= 0.9 + 0.09 + ...$$ (85b)

R and its Subspaces*

We add the following results to the information we now have about the various subspaces of **R*** to provide a full picture of the structure of the new real number system. The next theorem is a definitive result about the continuum **R***.

Theorem

In the lexicographic ordering **R*** consists of adjacent predecessor-successor pairs (each joined by d*); hence, the g-closure **R*** of **R** is a continuum [21].

Proof

For each N, N = 0, 1, ..., consider the set of decimals with integral part N. Take any decimal in the set, say, $N.a_1a_2...$, and another decimal in it. Without loss of generality, let $N.a_1a_2...$ be fixed and let it be the larger decimal. We take the average of the nth g-terms of $N.a_1a_2...$ and the second decimal; then take the average of the nth g-terms of this average and $N.a_1a_2...$; continue. We obtain the d-sequence with nth d-term, $(0.5)^{-n}a_1a_2...a_{n+k}$, which is a d-sequence of d*. Therefore, the g-limit of this sequence of averages is the predecessor of $N.a_1a_2...$ and we have proved that this g-limit and $N.a_1a_2...$ are predecessor-successor pair, differ by d* and form a continuum. Since the choice of $N.a_1a_2...$ is arbitrary then by taking the union of these predecessor-successor pairs of decimals in **R*** (each joined by the continuum d*) for all integral parts N, N = 0, 1, ..., we establish that **R*** is a continuum. □

However, the decimals form countably infinite discrete subspace of **R*** since there is a scheme for labeling them by integers. We can imagine them as forming a right triangle with one edge horizontal and the vertical one extending without bounds. The integral parts are lined up on the vertical edge and joined together by their branching digits between the hypotenuse and the horizontal that extend to d* which is adjacent to 0 (*i.e.*, differs from 0 by a dark number) at the vertex of the horizontal edge.

Corollary

R* is non-Archimedean and non-Hausdorff in both the standard and the g-norm and the subspace of decimals are countably infinite, hence, discrete but Archimedean and Hausdorff.

The following theorem that extends a theorem in Chapter 1 is standard in the real number system with the standard norm and is also true in the subspace of decimals. Therefore, we do not bring in d* in the proof so that this is really a theorem about the decimals with the standard norm which is not true in the g-norm because the decimals merge into a continuum at their tail digits and cannot be separated.

Theorem

Every real number is isolated from the rest.

Proof

Let p ∈ R be any irrational number and $\{q_n\}$ a sequence of rationals converging to p from the left. Let d_n be the distance from q_n to p and take an open ball of radius $d_n/10^n$, with center at q_n. Note that q_n tends to p but distinct from it for any n. Take an open ball of radius $d_n/10^n$, centered at p and take the union of open balls, centered at q_n, as n → ∞ and call it U. If r is any real, rational or irrational, to the left of p, then r is separated from p by two disjoint open balls, one in U and the other in its complement, center at p. If p is rational, then we take $\{q_n\}$ as a sequence of irrationals that tend to p, which is allowed by the Axiom of Choice. The same result will hold for any r distinct from and to the right of p [26].

This theorem shows that an irrational is not the limit of a sequence of rationals in the standard norm. Here is another shocker from the real number system:

Theorem

The rationals and irrationals are separated, *i.e.*, they are not dense in their union (this is the first indication of discreteness of the decimals).

Proof

Let p ∈ **R** (the real numbers including the ambiguous irrationals with the standard norm) be an irrational number and let q_n, n = 1, 2, ..., be a sequence of rationals towards and left of p, *i.e.*, n > m implies $q_n > q_m$; let d_n be the distance from q_n to p and take an open ball of radius $d_n/10^n$, center at q_n. Note that q_n tends to p but distinct from it for any n. Let U = ∪ U_n, as n → ∞, then U is open and if q is any real number, rational or irrational to the left of p then q is separated from p by disjoint open balls, one in U and, center at q and the other in the complement of U, center at p. Since the rationals are countable the union of open sets U for all the rationals and the irrational p is separated from all the rationals.

We use the same argument if p were rational and since the *reals* has countable basis we take q_n an irrational number, for each n, at center of open ball of radius $d_n/10^n$. Take U to be the union of such open balls. Using the same argument a real number in U, rational or irrational, is separated by disjoint open balls from p.

Thus, every decimal is separated from the rest of **R**, the terminating decimals from the nonterminating decimals and from each other. Clearly, the last two theorems do not hold in **R***. We restate another earlier theorem in Chapter 1 this time involving d* so that the theorem is true in **R***.

Theorem

The largest and smallest elements of the open interval (0, 1) are 0.99... and d*, respectively [26].

Proof

Let C_n be the nth term of the g-sequence of 0.99... For each n, let I_n be open segment (segment that excludes its endpoints) of radius 10^{-2n} centered at C_n. Since C_n lies in I_n for each n, C_n lies in (0, 1) as n increases indefinitely. Therefore, the decimal 0.99... lies in the open interval (0, 1) and never reaches 1. To prove that 0.99... is the largest decimal in the open interval (0, 1) let x be any point in (0, 1). Then x is less than 1. Since C_n is steadily increasing n can be chosen large enough so that x is less than C_n and this is so for all subsequent values of n. Therefore, x is less than 0.99... and since x is any decimal in the open interval (0, 1) then 0.99... is, indeed, the largest decimal in the interval and is itself less than 1.

To prove that 1 − 0.99... is the smallest element of **R**, we note that the g-sequence of 1 − 0.99... in (16) is steadily decreasing. Let K_n be the nth term of its g-sequence. For each n, let B_n be an open interval with radius 10^{-2n} centered at k_n. Then K_n lies in B_n for each n and all the B_ns lie in the open set in (0, 1). If y is any point of (0, 1), then y is greater than 0 and since the generating sequence 1 − 0.99... is steadily decreasing n can be chosen large enough such that y is greater than K_n and this is so for all subsequent values of n. Therefore, y is greater than = 1 − 0.99... and since the choice of y is arbitrary, 1 − 0.9... is the smallest number in the open interval (0, 1); at the same time 1 − 0.99... is greater than 0.

This theorem is true in **R** with d* replaced by 1 – 0.99… and follows from properties of terminating decimals but it was not known since neither 0.99… nor 1 – 0.99… was well-defined; it was assumed all along that 1 = 0.99… although the right side was ill-defined. The next theorem used to be called Goldbach's conjecture [9].

Theorem

An even number greater than 2 is sum of two primes.

The original proof of this theorem is in [23] during the early development of the new real number system but we shall reproduce it here for completeness. This is a conjecture in the real number system because, like Fermat's equation (FLT) [17], it is indeterminate. Before proving the theorem, we first note that an integer is a prime if it leaves a positive remainder when divided by another integer other than 1. We retain this definition in **R***.

Proof

The conjecture is obvious for n < 10. Let n be even greater than 10, p, q integers and p prime. If q is prime the theorem is proved; otherwise, it is divisible by an integer other than 1 and q. Since d* cannot be separated from any decimal, dividing q by an integer other than 1 and q leaves the remainder d* > 0. Therefore, q is prime. □

We now have a sense of how the decimals are arranged by the lexicographic ordering. Consider the decimals with integral part N:

$$N.99\ldots\ldots\ldots\ldots$$

$$N.4800\ldots\ldots\ldots$$

$$N.4799\ldots\ldots\ldots$$

$$N.10\ldots\ldots\ldots\ldots..$$

$$N.00\ldots0100\ldots$$

$$N.00\ldots. \tag{86}$$

The largest decimal in the set is the decimal integer N.99… and the smallest is the terminating decimal N.00… [26]. Starting from the bottom going up, the decimals with integral part N are arranged as predecessor-successor pairs each joined by d*. Each gap (ellipses) is filled by countably infinite adjacent predecessor-successor pairs also joined by d* so that their union is a continuum. We now have a clear picture of how **R*** is arranged in the new real line linearly ordered by <, the lexicographic ordering.

Important Results; Resolution of a Paradox

(1) Every convergent sequence has a g-subsequence that defines a decimal adjacent to its limit in the standard norm. If the decimal is terminating it is the limit itself.

(2) It follows from (1) that the limit of a sequence of terminating decimals can be found by evaluating the g-limit of its g-subsequence which is adjacent to it. We can use this as alternative way of computing the limit of ordinary sequence.

(3) In [72] several counterexamples to the generalized Jourdan curve theorem for n-sphere [71] are shown where a continuous curve has points in both the interior and exterior of the n-sphere, n = 2, 3,…, without crossing the n-sphere [71, 72]. The explanation is: the functions cross the n-sphere through dark numbers.

(4) Given two decimals and their g-sequences and respective nth g-terms A_n, B_n we define the nth g-distance as the g-norm $|A_n - B_n|$ of the difference between their nth g-terms. Their g-distance is the g-lim $|A_n - B_n|$, as n → ∞, which is adjacent to the standard norm of the difference [21]. Advantage: the g-distance is the g-norm of their decimal difference; the difference between nonterminating decimals

cannot be evaluated otherwise. Moreover, this notion of distance can be extended to n-space, n – 2, 3,..., and the distance between two points can be evaluated digit by digit in terms of their components without the need for evaluating roots. In fact, any computation in the g-norm yields the results directly, digit by digit, without the need for intermediate computation such as evaluation of roots in standard computation. (The decimals are glued together by d* to form **R***)

More on Nonstandard Numbers

We highlight some properties of a special class of nonstandard numbers that can be checked by looking at their g- or divergent sequences. The principal element of d* (g-limit of its principal g-sequence) is dark number of order 0.1.

Let γ be a fraction such that $0 < \gamma < 1$ and let $d_\gamma = \text{g-lim}\gamma^n$, as $n \to \infty$, n integer, d_γ is called dark number of order γ. An unbounded number u of order $\lambda > 1$ is defined as the upper bound of the sequence

λ^n, as $n \to \infty$. The number u is an element of u* just as d_γ is an element of d*. Since γ^n is positive and steadily decreasing, d_γ is less than any given decimal. (In this section we only consider positive decimal and hence we shall drop the qualification positive) To see this, let x be any decimal; since $\gamma < 1$, the integer n can be chosen large enough that $0 < d_\gamma < \lambda^n < x$. Similarly, it can be shown that an unbounded decimal of any order is greater than any decimal.

The following is obvious by checking their g-sequences.

1.   The product of any nonzero decimal and dark number of order γ is dark number of order γ; the product of a decimal and unbounded number of order λ is unbounded of order λ.

2.   If d_1 and d_2 are dark numbers of order γ_1 and γ_2, respectively, where $\gamma_1 < \gamma_2$, then $d_1 + d_2$ is dark number of order γ_2, $d_2 - d_1$ is dark number of order γ_1, d_1/d_2 is dark number of order γ_1/γ_2 and d_1/d_2 is unbounded number of order γ_2/γ_1.

3.   A decimal divided by a dark number of order γ is unbounded of order $1/\gamma$; a decimal divided by unbounded number of order λ is a dark number of order $1/\lambda$; the reciprocal of dark number of order γ is unbounded number of order $1/\gamma$; the reciprocal of unbounded number of order λ is dark number of order $1/\lambda$.

4.   If μ_1, μ_2 are unbounded numbers of orders λ_1, λ_2, respectively, where $\lambda_1 > \lambda_2$, then $\mu_1 + \mu_2$ and $\mu_1 - \mu_2$ are both unbounded numbers of order λ_1 and μ_1/μ_2 and μ_2/μ_1 are unbounded and dark numbers of orders λ_1/λ_2 and λ_2/λ_1, respectively.

5.   The sum of two dark numbers of the same order is a dark number of that order; the quotient of two dark numbers is indeterminate but can be avoided using nth g-term approximation. If the nth g-term of the quotient is a decimal then the quotient is a decimal, if it is greater than 1 the quotient is u*; if it is less than 1 the quotient is d*.

These results, taken from [16, 21], are useful in avoiding indeterminate forms in calculation. Moreover, since all elements of d* share the properties of d*, we can use any element of this class for our argument in proving a theorem, especially, in dealing with inequality, the advantage being that it has clear structure. Consequently, there is no loss of generality in using the principal nth g-term of d* for any purpose involving d*.

The Counterexamples to FLT

Given the contradiction in negative statement, we use Fermat's equation in place of Fermat's last theorem (FLT) so that its solutions are counterexamples to FLT. We first summarize the properties of the digit or basic integer 9.

(1)   A string of 9s differs from the nearest power of 10 by 1, e.g., $10^{100} - 99...9 = 1$.

(2)   If N is an integer, then $(0.99...)^N = 0.99...$ and, naturally, both sides of the equation have the same g-sequence. Therefore, for any integer N, $((0.99,..)10)^N = (9.99...)10^N$.

(3) $(d^*)^N = d^*$; $((0.99,..)10)^N + d^* = 10^N$, N = 1, 2, ...

Then the exact solutions of Fermat's equation are given by the triple $(x, y, z) = ((0.99...)10^T, d^*, 10^T)$, T = 1, 2, ..., that clearly satisfies Fermat's equation,

$$x^n + y^n = z^n, \tag{87}$$

for n = NT > 2. Moreover, for k = 1, 2, ..., the triple (kx, ky, kz) also satisfies Fermat's equation. They are the countably infinite counterexamples to FLT that prove the conjecture false [17]. One counterexample is, of course, sufficient to disprove a conjecture.

Introduction to Discrete Computation and Calculus

Discrete computation is particularly suitable for DEVS and simulation [120, 121].

Well Behaved Functions

We call well-behaved polynomial, rational, exponential, logarithmic and circular functions as well as their sums, products, quotients and composites away from points of discontinuity. We consider well-behaved functions on the terminating decimals **R** which are discrete but continuous on **R*** the latter being a continuum.

Since **R** is discrete its image under a well-behaved function is discrete. However, since there are dark numbers between points in its graph the latter appear continuous, the gap only of interest for computing and applications, especially, simulation where the tools are discrete. Then there is no need to approximate continuous function by discrete function as done in [120, 121].

Computation

Computation is mapping of function or algebraic operation on functions into a number system or system of functions. Computation includes finding the result of algebraic operations on functions and evaluation of their values and limits.

We recall some of the concepts involved in computation. One is the limit point of topology (we simply refer to it as limit) in the standard norm which is clearly defined. The point P is limit of a sequence or series (sum of terms of a sequence) if every neighborhood of P contains a term of the sequence or series. What is its relationship to the g-norm? They are adjacent (differ by d*) because the standard norm of a decimal $N.a_1a_2...a_n...$ is the sum of its series expansion,

$$N.a_1a_2...a_n... = N +.a_1 +.0a_2 + ... +.0...0\ a_n...$$

$$= N + \Sigma(1)^{-n}a_n, n = 1, 2, ... \tag{88}$$

This sum in the standard norm is adjacent to the decimal; therefore, it approximates the decimal by an error of d*.

Since the g-norm is precise, *i.e.*, yields the result of computation directly as a decimal digit by digit, the margin of precision is determined by the number of decimal digits computed; the intermediate steps of standard computation, e.g., evaluation of roots, are avoided. Then the limit is approximated by the g-limit, *i.e.*, the decimal, to any level of accuracy within 10^{-n}, the nth d-term of d*. Thus, computation by the g-norm saves considerable computer time.

As example, we note that the standard norm or magnitude of the nonterminating decimal 0.99... is the sum of the series,

$$0.99... = 0.9 + 0.09 + ... 0.00...09 + ... = \Sigma 9(1)^{-n}, \tag{89}$$

which is an approximation of 1 at margin of error d* and 1 is adjacent to its g-norm, 0.99..., since $1 - 0.99... = d*$ [26]. Of course, d* does not show in (86) being dark but $d* + 0.99... = 1$.

For purposes of computation we denote the nth g-term of a decimal by the functional notation n-ξ(x) called n-truncation. Since a g-sequence defines or generates a decimal we call the latter its g-limit. Since nonterminating decimals cannot be added, subtracted, and divided, they must be n-truncated first to carry out the operations on them. The margin of error at each step in the computation must be consistent (analogous to the requirement of number of significant figures in physics, the rationale being that the result of computation cannot be more accurate than any of the approximations of the terms). While we can start division by terminating decimal on the left digits the quotient is nonterminating when the divisor has a prime factor other than 2 or 5.

Let $x = N.a_1\ldots a_n\ldots$ and $y = M.b_1\ldots b_n\ldots$, then

$$n\text{-}\xi(x) = N.a_1\ldots a_n, \quad n\text{-}\xi(y) = M.b_1\ldots b_n,$$

$$n\text{-}\xi(x + y) = n\text{-}\xi(x) + n\text{-}\xi(y),$$

$$n\text{-}\xi(x - y) = n\text{-}\xi(x) - n\text{-}\xi(y),$$

$$n\text{-}\xi(xy) = (n\text{-}\xi(x))(n\text{-}\xi(y)),$$

$$n\text{-}\xi(x/y) = (n\text{-}\xi(x))/(n\text{-}\xi(y)), \tag{90}$$

provided $n\text{-}\xi(y) \neq 0$ as divisor. Consider the function $f(x_1, \ldots, x_k)$ of several variables; we n-truncate f as follows:

$$n\text{-}\xi(f(x_1, \ldots, x_k)) = f(n\text{-}\xi(x_1), \ldots, n\text{-}\xi(x_k)). \tag{91}$$

If f is a composite function of several variables, $f(g_1(x_1, \ldots, x_t), \ldots, g_s(y_1, \ldots, y_u))$ then,

$$n\text{-}\xi(f(g_1(x_1, \ldots, x_t), \ldots, g_s(y_1, \ldots, y_u)))$$

$$= (n\text{-}\xi(g_1(n\text{-}\xi(x_1)), \ldots, n\text{-}\xi(x_t)), \ldots, n\text{-}\xi(g_s(n\text{-}\xi(y_1)), \ldots, n\text{-}\xi(y_u))). \tag{92}$$

This formalizes standard computation in the new real numbers. The computation itself uses the g-terms of the decimals involved and provides the result directly, digit by digit; it approximates the result to within any d-term of d*, the closest approximation one can ever get to is d* as in (87). Computation using the g-norm applies to monotone increasing function since the g-terms of a decimal is monotone increasing. However, a monotone decreasing function can be converted to a monotone increasing one and g-norm computation applied to the latter.

We give simple examples below to illustrate the methodology without getting distracted by unnecessary complexity [21]. Consider the monotone increasing function,

$$f(x) = x^{1/3}. \tag{93}$$

We want to evaluate f(5) to within 3 decimal digits. We make a series of 3-trunctions of f(5) to find the first three g-terms of its g-sequence.

Step 1. Find the largest integer N such that $N^3 \leq 5$. Clearly, N = 1.

Step 2. Divide segment [0, 1] by points, 0, 0.1, 0.2, …, 0.9, 1, and find the largest number a_1 such that $(1.a_1)^3 \leq 5$. If $(1.a_1)^3 = 5$, f(5) is a terminating decimal. This is not so here since, $a_1 = 0.7$ and $(1.7)^3 = 4.913$ and the first term of the g-sequence is 1.7.

Step 3. Divide the segment [0, 0.1] by the points 0, 0.01, 0.02, …, 0.09, 0.1] and find the largest number among them and call it a_2 such that $(1.7a_2)^3 \leq 5$. In this case, $a_2 = 0$. Then the first three digits of f(5) are known: 1.70.

Step 4. Find a_3 such that $(1.70a_3)^3 \leq 5$; then $a_3 = 9$ and $(1.709)^3 = 4.991$. Thus, the 3rd g-term of f(5) = 1.709. The calculation can be carried out to find the nth term of the g-sequence of f(5) for any n. Actually, we calculated the first three terms of the g-sequence of $x^{1/3}$ or its 3rd g-term, the way the scientific calculator computes cube root.

This calculation applies to any well-behaved function since every point in it except point of discontinuity has a neighborhood in which it is monotone. We used the g-norm to compute the result directly digit by digit. This is exactly how the calculator does it. It may look time-consuming but it can be done in split second with the right software.

Suppose we have the composite function, $h(x) = f(g(x))$, where $f(x) = x^{1/2}$, $g(x) = x + 1$ so that $f(g(x)) = (x + 1)^{1/2}$. We want to evaluate $h(9)$ up to the 3rd decimal digit, *i.e.*, at 10^{-3} margin of error.

Step 1. We want to find the 3-truncation $N.a_1a_2a_3$ of $h(9)$. We first compute the integral part. Obviously, the largest integer N such that N^2 does not exceed $(h(9))^2 = 10$ is $N = 3$.

Step 2. Divide the interval [0, 1] by the points, 0.1, 0.2, 0.3, 0.4, 0.5, 0.6, 0.7, 0.8, 0.9, 1, and find the largest among them and denote it by $0.a_1$ such that $(3.a_1)^2$ does not exceed 10. that decimal is 0.1.

Step 3. Divide the interval [0, 0.1] by the division points 0.01, 0.02, 0.03, 0.04, 0.05, 0.06, 0.07, 0.08, 0.09, 0.1 and find the largest of the division points and denote it by $0.0a_2$ such that $(3.1a_2)^2$ does not exceed 10. The decimal is 0.06.

Step 4. Divide the interval [0, 0.01] by the division points 0.001, 0.002, 0.003, 0.004, 0.005, 0.006, 0.007, 0.008, 0.009, 0.01 and find the largest among the division points and denote it by $0.00a_3$ such that $(3.16a_3)^2$ does not exceed 10. That number is 0.002.

Therefore, the 3rd g-term approximation of $h(9)$ at margin of error $(10)^{-3}$ is 3.162.

To find fractional root of a decimal x, say, $h(x) = x^{k/m}$, we look it as composite function $f(g(x))$, where $f(x) = x^{1/m}$, $g(x) = x^k$ both of which are monotone increasing. Then we can n-truncate each of $f(x)$ and $g(x)$ and then n-truncate the corresponding composite of their n-truncations to obtain the nth g-term approximation of the composite function $h(x)$. Extension to sum, product and quotients is obvious.

The Limit of a Sequence of Decimals

We recall that point P is the limit of the g-sequence or series expansion of a nonterminating decimal if every open interval containing P contains an element of the sequence or series. For example, 1 is the unique limit of 0.99... Since the limit and g-limit of a nonterminating decimal are adjacent the g-limit of the value of a function approximates the value at margin of error d* or the nth g-term at margin of error 10^{-n}. The terminating decimal 4.5300... and 4.5299... are adjacent the former being the successor of the latter in the lexicographic ordering of the decimals. Therefore, the former is the limit of the g-sequence of the latter. A nonstandard number, aside from d* and u*, is the sum of a decimal and d*. This means that its g-sequence has a set of digits that moves to the right indefinitely leaving fixed digits behind. The digits that move to the right are the nonstandard nth d-terms of d* and the digits that remain fixed are the digits of the g-terms of the decimal. Since d* cannot be separated from the decimal as its dark component except that we cannot identify its dark component being a point in the continuum we can look at the latter as the standard component. Moreover, d* cannot be separated from 0 although distinct from it; therefore, it is adjacent to it.

The sequence, 1.25315, 1.250153, 1.2500351, 1.25000531, ... shows the g-terms of some nonterminating decimal, and the receding d-terms of d* so that the nonstandard decimal is 1.25 + d* that reduces to the standard decimal 1.25. In other words, we can look at a decimal as approximation of some nonstandard number and the margin of error is d*. What is the purpose of all these?

Consider the function,

$$F(x) = H(x) + \delta(x), \tag{94}$$

where $H(x)$ does not diverge as x tends to some limit and $\delta(x)$ tends to 0 as limit; then in the calculation of the limit of $F(x)$ as $x \to s$ the nth g-term will consist of two components one with digits remaining fixed and another set of digits that recedes indefinitely to the right, the nth d-term of d*. Thus, in evaluating limit of a function we keep computing the value of the function (iterated computation) over finer and finer refinements of the sequence of termininating decimals that tends to s as limit (there is no loss of generality in taking successive averages between the terms of the sequence and s). We call H(x) and δ(x) the principal and minor parts of F(x), respectively. Even if the function is not separated into principal and minor parts, they will show in the calculation of its limit. Moreover, since the computation involves iterated approximations the problem of indeterminacy does not arise.

If the function is a sequence of terminating decimals $\{a_n\}$, n = 1, 2, ..., we take the values of the sequence along n, which need not be consecutive values. To facilitate convergence we may skip some values of n and take large values.

Consider our previous example, the sequence of numbers, $f(n) = (n^4 + 1)/n^4$. We compute the terms along n = 1, 2, ..., and note their truncated sequence of values:

n = 1, f(1) = 2.0000000

n = 2, f(2) = 1.0625000

n = 3, f(3) = 1.0123456

n = 4, f(4) = 1.0039062

n = 50, f(50) = 1.0000000.   (95)

Discarding the first term f(1) = 2.0000000 corresponding to n = 1, which is not a g-term, we have the nonstandard g-sequence of f(n),

1.0625000, 1.0625000, 1.0123456, 1.0039062,

1.0016000, ..., 1.0001000, ...,   (96)

whose limit is 1 + d*. Note the nth d-term with set of digits varying and receding to the right indefinitely leaving the fixed digits behind. The varying elements are the d-terms of d*. The fixed digits left behind are the g-terms of 1, a terminating decimal.

Consider the limit of the sequence, $f(n) = (2n^4+1)/3n^4$, n = 1, 2, ... We find a nonterminating decimal that is adjacent to it. We do the following computation:

n = 2, f(2) = 0.6875000, n = 6, f(6) = 0.66669238

n = 3, f(3) = 0.6707818, n = 8, f(8) = 0.66674800

n = 4, f(4) = 0.6679687, n = 7, f(7) = 0.66680540

n = 5, f(5) = 0.6672000, n = 9, f(9) = 0.66671740

n = 100, f(100) = 0.66666666.   (97)

Thus, the nonterminating decimal adjacent to the limit of f(n), as x → ∞, has periodic initial segment.

We now compute the g-limit of a function. Let f(x) be a function and suppose we want to find limf(x), as x → s. We find limf(x) along successive refinements starting with the steadily increasing sequence $x_0, x_1, x_2, ...,$ left of and towards s > 0 as limit. We refine the sequence by inserting the succession of averages of s and x_0, of s and x_1, etc., ..., relabeling them as $s_1, s_2, s_3, ...,$ etc. so that the refinement becomes the sequence, $x_0, s_1, s_2, s_3, ...$ We continue the refinement and compute the values of the nth g-term of g-limf(x) along the kth refinement. If for suitable value of k we find a set of digits in the nth g-terms of f(x) that recedes to the right leaving fixed digits behind up to the nth term, that would be the nth g-term of the g-limit which is adjacent to the limit. If *s* is nonterminating, we obviously need to truncate *s* to desired accuracy for the computation. The advantages of this scheme in computing f(x) or limf(x) for given x is quite clear. For instance, a program for quickly computing limit of well-behaved function can be developed to obtain the result in split second. This is not possible for non-well-behaved functions such as wild oscillation [19, 20].

Computation with Nonstandard Numbers

Consider the function f(x) = g(x) + d(x) in the neighborhood of a decimal s, where x is decimal and g(x) (principal part) and d(x) (minor part) are decimal-valued functions and g(x) tends to a nonzero decimal and d(x) tends to 0, as x → s. Then g-lim f(x) = g-lim (g(x) + g-lim b(x)) = g-lim g(x), as x → a. If g-lim g(x) is unbounded at s then g-lim d(x) is unbounded. In computation we treat nonstandard function f(x) as binomial, the sum of principal and minor parts. In algebraic operations involving sum and product we write the result in the form F(x) = G(x) + η(x), where F(x) and η(x) are the principal and minor parts, respectively. Then g-limF(x) = g-limG(x), as x → s. The minor part of the function G(x) may be discarded in calculating its g-limit; if they appear as factors, the arithmetic of dark numbers applies.

If λ, $0 < \lambda < 1$, is a terminating decimal, then the g-limit of the sequence λ^n, n = 1, 2, ..., is d* and λ is called dark number of order λ.

Consider the non-uniformly bounded convergent sequence S [26]:

$$0.123, (0.312)^2, (0.231)^2, (0.123)^3, ...,\qquad\qquad\qquad\textbf{(98)}$$

whose terms are cyclic permutations of the digits 1, 2, 3. To find its limits we split it into three component sequences:

(a) $0.123, (0.123)^2, (0.123)^3, ...,$

(b) $0.312), (0.312)^2, (0.312)^3, ...,$

(c) $0.231, (0.231)^2, (0.231)^3, ...,\qquad\qquad\qquad\qquad\textbf{(99)}$

which converge to distinct elements of d*. Therefore, the g-limit of (97) is three-valued, consisting of dark numbers of these orders. Since the number of ways of forming such component sequences is countable one may form a dark number of countable set-valued order as well as nonstandard functions with set-valued principal and minor parts. Since they all recede to the right indefinitely they become indistinguishable and their d limit is a continuum.

A module $H(x_1,..., x_k)$ is a rational expression in the variables $x_1, ..., x_k$. Suppose the values of the arguments are given and H is computable as a single decimal, terminating or nonterminating. Then the value of H can be computed to any margin of error. If this is not the case, then compute the nth g-terms of the arguments, at consistent margin of error, and find the value of H in terms of those nth g-terms. Given two modules H and G and using the approximation function ξ_n, we define $\xi_n(H + G) = \xi_n(H) + \xi_n(G)$ and $\xi_n(HG) = \xi_n(H)_n\xi_n(G)$ at consistent margin of error. In dealing with nonterminating decimals it may not be possible to verify equality between two modules by actual computation. In this case, we say that two modules $H(x_1, ..., x_k)$, $G(x_1, ..., x_m)$ are equal if their n-truncations are equal, *i.e.*, $\xi_n(H) = \xi_n(G)$, for each n = 1, 2, ...

Although dark numbers of some orders are special elements of d* they share the properties of d* and we may substitute d* in any equation or expression involving them.

Consider the function $H(x) = h(x) + d(x)$, where $h(x)$ tends to a decimal and $d(x)$ tends to 0, as $x \to s$. Then g-lim$H(x)$ = g-lim$h(x)$ since the sum of a decimal and dark number is the same decimal. If some function $\nabla g(x)$ is small and tends to 0 as limit, *i.e.*, dark number of some order γ, and lim$h(x)$ is a decimal then lim$(\nabla g(x)H(x))$, $x \to s$, is dark number of order γ provided $h(x)$ does not diverge. A function of the form $\nabla g(x)hx)$, as $x \to s$, is dark number of order α if $\nabla g(x)$ tends to the form α^n, where $n \to \infty$, as $x \to s$. Note that g-lim$H(x)$ = g-lim$h(x)$ + g-lim$d(x)$ = g-lim$h(x)$. Thus, taking the g-limit of a function amounts to discarding the minor part of the nonstandard function provided the principal part does not diverge. If the g-limit is terminating decimal it is generally obtained by substitution because it is actually attained. For example, if $f(x)$ is the principal part of some nonstandard function, say, $H(x) = x^2 + d(x)$, then g-lim$H(x)$, as $x \to a$, is a^2. The sum or product of nonstandard functions is obtained by considering each function as binomial, the sum of its principal and minor parts. This is also the way to handle nonstandard numbers and operations and inverse operations. If the divisor tends to 0 then it is a divergent sequence of u^* and the arithmetic of u^* applies. For instance, if the quotient has the form $d_1(x)/d_2(x)$ and numerator and denominator tend to 0, *i.e.*, dark numbers of orders γ_1, γ_2, respectively, the quotient has order $\gamma = \gamma_1/\gamma_2$ and is either a dark number, decimal or unbounded number depending on whether $\gamma < 1$, $\gamma = 1$ or $\gamma > 1$. However, by n-truncation to find the nth g-terms, we can tell what the g-limit would be and so indeterminacy is avoided. Moreover, for finding limit of ordinary indeterminate form, truncation will compute the g-limit directly without being bothered by indeterminacy.

Discrete Optimization

If a function is well-behaved in the neighborhood of a point P there is a neighborhood of P at which the function is steadily increasing or decreasing; if the function is steadily increasing it has a maximum M. If we want to find its minimum we find some constant K so that $K - f(x)$ is steadily increasing. Then max$(K - f(x))$ = min$(f(x))$. If the maximum M exists, maximum of $(M - f(x))$ = minimum of $f(x)$ also exists. Note that this algorithm for computing maximum or minimum yields the answer directly as a decimal which is adjacent to the limit (standard norm). This offers some advantage over the standard norm, especially, in computation that requires finding the roots. This is avoided because the digits of the result are obtained directly by n-truncation as in the above examples. We illustrate optimization with a simple function along continuous refinements of division points in an interval on its domain that highlights the methodology. Consider the function $y = f(x) = 3x - x^2 - 2$. We start with the values of $f(x)$ on the interval between its zeros, $x = 1$ and $x = 2$. We find its values along midpoints (averages) until the values taper off to a certain value. Then we have the following values of $f(x)$:

x |1.0000|1.0625|1.1250|1.1875|1.2500|1.3125|1.3750|1.4375|

f(x) |0.0000|0.0586|0.1094|0.1523|01.875|0.2148|0.2344|0.2461|

x |1.5000|1.5000|1.5625|1.6250|1.6875|1.7500|1.8123|1.8750|

f(x) |0.2500|0.2461|0.2344|0.2344|0.2148|0.1875|0.1525|0.1094⌋

f(2) = 0   **(100)**

The value of the function along refinements tapers off at 0.2500 and if we continue along the relevant terms of the sequence of values of x that tends towards $x = 1$ we will generate the decimal 2.4999... which is adjacent to 1, the maximum of $f(x)$ and its g-norm at margin of error 10^{-4}. Moreover, symmetry can be detected from the table above. The use of averages or midpoints is a matter of convenience. Any sequence of refinements will do (see extension to discrete functions in [21]).

Advantages of the g-Norm

The advantages of the g-norm over other norms are as follows:

(a) It avoids indeterminate forms.

(b) Since the g-norm of a decimal is itself, computation with it yields the answer directly as a decimal, digit by digit, and avoids intermediate approximations of standard computation. It means considerable reduction in time for large computations.

(c) Since the standard limit is adjacent to the g-limit of some g-sequence, evaluating it reduces to finding some nonterminating decimal adjacent to it; the decimal is approximated by the appropriate nth g-term. Both the computation and approximation are precise. In fact, the exact margin of error is d*. This applies to the result of any computation: it is adjacent to some nonterminating decimal and the latter is found using the g-norm.

(d) In iterated computation along successive refinements of sequence x_j that tends to a as $j \rightarrow \infty$, the iteration is simplified by taking midpoints or averages between the sequence of points x_j and the g-limit s.

(e) Approximation by nth g-term or n-truncation contains the ambiguity of nonterminating decimals.

(f) Calculation of distance between two decimals is direct, digit by digit, and requires no square root. In fact, calculation by the g-norm involves no root or radical at all.

(g) In general radicals in computation, e.g., taking root of a prime, is avoided by nth g-term approximation or n-truncation to any desired margin of error where accuracy is measured by the number of digits of the result obtained.

To summarize the g-norm is the natural norm for purposes of computation for it does three things: (a) it puts rigor in computation since every step is in accordance with the new definition of the previously ill-defined nonterminating decimals (meaningless infinite arrays of digits most of which unknown) in terms of the well-defined terminating decimals, (b) the margin of error is precisely determined and (c) the result of computation is obtained digit by digit and avoids intermediate unnecessary approximations of standard computation. For large computation (c) provides significant reduction in computing time as it avoids intermediate approximations and proceeds directly to the calculation of the digits of the result.

Introduction to Discrete Calculus

With the g-norm we set up the mechanism for discrete differentiation and integration since both involve limits. This is all we can do here: some concepts of discrete differentiation and integration.

To find the derivative at x = s, where s is terminating decimal, we find the nth g-term of $\Delta f / \Delta x$, as $\Delta x \rightarrow 0$, i, e., $x \rightarrow s$, where $\Delta x = x - s$, starting from a point x_0 near s where $\Delta f / \Delta x$ is steadily increasing or decreasing along successive points x_0, x_1, etc., and where $\Delta_0 = x_0 - s$, $\Delta_1 = x_1 - s$, etc. This way, we find the g-sequence of the g-limit (the exception is when $\Delta f / \Delta x$ diverges). If the derivative of f(x) exists in the sense of calculus it also exists in discrete calculus and the differentiation rules of calculus applies. We have here a simpler technique for optimization. Moreover, even function having no derivative in calculus such as set-valued function or function having set-valued derivative may have an optimum. Ref. [19, 20] discusses set-valued *functions* and derivatives including wild oscillation of the form $f(x) = \sin^n(1/x^2)$, n = 1, 2, ... When Δx tends to 0 in the standard norm but Δf tends to a nonzero decimal then $\Delta f / \Delta x$ diverges.

The extension of our computational technique to composite function of several functions of several variables is straight forward and similar to the techniques of calculus. The only innovation here is the use of truncation for finding g-sequences.

The graphs of well-behaved functions in $\mathbf{R}^* \times \mathbf{R}^*$ and $\mathbf{R} \times \mathbf{R}$ are the same since each missing element is the dark number d* squeezed between adjacent decimals which is not detectable.

The functions of discrete calculus are functions over the decimals. Therefore, they are discrete-valued. Consider the function y = f(x) in the interval [a, b] and subdivide [a, b] by the finite set of points (decimals) $\{x_k\}$, k = 1, 2, ..., s, where $a = x_1$, $b = x_s$. We further subdivide the set by the finite set of points $\{x_m\}$, m = 1, 2, ..., t, take $\{x_k\} \cup \{x_m\} = \{x_n\}$, n = 1, ..., w, w = s + t, and call it $\{x_n\}$, n = 1, ..., w, where we relabel the points of the union of the two sets, preserving their lexicographic ordering and taking $a = x_1$, $b = x_t$, a refinement of both $\{x_k\}$ and $\{x_m\}$. For a well-behaved function, except at points where it is undefined, there is no significant difference for purposes of evaluating its value or limit in taking midpoints of the subsegments determined by each refinement of the subdivision of the interval [a, b].

One advantage of discrete function is: we need not differentiate it to find a maximum or minimum; naturally, problems of optimization unsolvable in calculus may be solved here. This technique applies to some functions with set-valued derivatives such as the schizoid or curve with cusp. It is also applicable to set-valued functions.

However, the case where the given function is wild oscillation such as $f(x) = \sin^n 1/x$, $n = 1, 2, ..., k$, requires special technique appropriate for set-valued functions [13].

Consider the arc of the function $y = f(x)$ over the interval [a, b] and let the set of points $\{x_k\}$, $k =, 1, 2, ..., s$, subdivide the interval [a, b] suitably so that each local maximum or minimum is isolated in some interval. This is possible since the values of $f(x)$ over the decimals are countably infinite and discrete. Unless the maximum or minimum lies at an end point, its neighborhood will contain interval on which $f(x)$ is increasing on one side and decreasing on the other. At the same time, the end point of a function in an interval is either minimum or maximum. In fact, every closed interval in the range of a function contains its maximum or minimum. Without loss of generality, consider interval [c, d] containing a maximum. Subdivide the interval [c, d] and take successive refinements $\{x_m\}$ of $\{x_k\}$ ($\{x_m\}$ relabeled suitably) until the values of $f(x_m)$, tapers to a constant α along $\{x_m\}m = 1$, $2, ..., w$. In this new setting the values of a function, being discrete, is no different from a sequence of numbers. As the values become close to each other they contain a d-sequence with a set of digits in its terms receding to the right and forming a d-sequence of d* and another set of values that remain fixed. The latter defines a decimal, a local maximum in this interval. By suitable translation of the function the minimum can be similarly obtained (the end point is either a maximum or minimum). Then the absolute maximum of $f(x)$ is the maximum of M_k, $k = 1, 2, ..., m$. In this algorithm for finding the g-limit of a function there will be, in general, several inequivalent g-sequences each a g-sequence of a local maximum. A single g-sequence may split into distinct g-sequences in further computation of the nth g-terms when their limits are close to each other. Some functions have countably infinite maxima, e.g., infinitesimal zigzag and oscillation [24, 34].

This method applies to function with maximum at cusp, e.g., schizoid. By suitable transformation the minimum can be found in a similar way. This approach is both intuitive and computational. A more sophisticated version of it for discrete function is approximation of continuous function developed in [120, 121].

We extend our method to the calculation of the length of an arc of a curve. Consider the function $y = f(x)$ over the interval [a, b]. Let the set of points $\{x_k\}$, $k = 1, 2, ..., s$, subdivide the interval [a, b] and form the sum,

$$L_k = \sum((x_{k+1} - x_k)^2 + (f(x_{k+1} - x_k)^2)^{1/2}, k = 1, 2, ..., s, \tag{101}$$

where $a = x_1$, $b = x_s$. Take refinements of $\{x_m\}$ of $\{x_k\}$ ($\{x_m\}$ relabelled suitably) until the values of L_m tapers to a constant Γ; we call Γ the length of the curve of $y = f(x)$ over [a, b].

In calculus the right derivative of a curve at point P is obtained by drawing a line from P to a point Q nearby and moving Q towards P from the right along the curve, the derivative being the limit of the slope of the line PQ as Q moves towards P (actually, Q moves along discrete points, *i.e.*, approximation of "continuous" curve by polygonal line). We find the right discrete derivative similarly by taking Q to move along discrete set of points, the advantage being that this is purely computational. The right discrete derivative is found in the same manner by taking the limit of the quotient $\Delta f(x)/\Delta x$, as $\Delta x \to 0^+$, along suitable refinement. When the right and left derivatives at a point are equal then we say that the curve is discretely regular there. To find the left discrete derivative of the function $y = f(x)$ at the point P(b, f(b)] on the interval [a, b], we assume $f(x) > 0$ and increasing. Subdivide the interval [a, b] by the set of points $\{x_k\}$, $k = 1, 2, ..., s$, where $a = x_1$, $b = x_s$, and form the quotient,

$$D_L(f(b)) = (f(x_{k+1} - x_k))/(x_{k+1} - x_k). \tag{102}$$

We take successive refinements of $\{x_k\}$ to obtain the g-sequence of a decimal or divergent sequence of u*. Then D_L is either terminating, nonterminating, zero or u*, taking approximation in each case. The right derivative D_R can be computed similarly. When the function is discretely regular at x we denote its discrete derivative at x by D_x.

We introduce the notion of locally approximating the shape of a steadily increasing function on [a, b] at a point by its derivative at P, as P traces the arc over [a, b], obtained by finding the limit of the minimum of the maximum horizontal distance between the derivative function and the function itself as P traces the arc [26]. Computation is straight forward for well-behaved functions. The limit gives the shape of the curve in a small neighborhood of P expressed by the following theorem for smooth curves that applies as well to discrete curves since the gaps are dark.

Minimax Principle

When the minimum of the maximum horizontal distance between two simple smooth arcs with no inflection point can be made arbitrarily small then an element of arc and variation of derivative at a point on one approximates the other [26].

We define the integral of f(x) over the interval [a, b] as the limit of the sums of the areas of the trapezoidal areas under the curve through successive refinements determined by the midpoints of the subsegments at each refinement, as we do in calculus except that it is simpler here since the upper and lower sums coincide. This is another advantage with discrete function. The indefinite integral of f(x) is simply the area under f(x) over the interval [a, x]. Note that the integral of f(x) is independent of the derivative. To evaluate it we divide the interval [a, b] into subintervals by the points $\{x_k\}$, k = 1, 2, ..., s, a = x_1, b = x_s, form the sum,

$$\sum_{[a, b]} f(x) = \sum (x_{k+1} - x_k)(f(x_{k+1}) + f(x_k))/2, \text{ from } k = 1 \text{ to } k = s, \tag{103}$$

and find its limit through successive refinements of $\{x_m\}$, as m→∞.

From these examples, we find that computation in the standard norm reduces to finding the nonterminating decimal adjacent to the result and, therefore, approximable to any desired margin of error.

This is just a framework for building discrete calculus appropriate for computing and simulation. Simulation is important for finding ballpark estimate of hidden forces (attractive or repulsive). For example, distortion of the orbit of a planet reveals the presence of some cosmological body whose gravity impinges on the planetary orbit. Then by simulating different masses and gravitational forces at different points in the neighborhood of the distortion one may get the best fit and take that sight as potential region for searching the unknown mass.

THE COMPLEX VECTOR PLANE

Consider this statement that, once raised, invariably sparks endless spirited debate:

$$i = \sqrt{(-1)} = \sqrt{(1/-1)} = 1/i = -i \text{ or } 2i = 0 \text{ or } i = 0. \tag{104a}$$

If in (104) we divide the equation i = −i by i or add −i on both sides, we obtain,

$$1 = -1 \text{ or } 1 = 0, 2 = 0, ..., i = 0 \text{ and } x = 0; \tag{104b}$$

for any real number x and both the complex and real number systems collapse. We found earlier that the vacuous concept i is to blame here.

The Number j as Operator on Plane Vectors

Our proposed remedy is to introduce an operator **j** in place of I (Appendix to [19]). We define **j** as left-right operator on or mapping of a plane vector by positive or counterclockwise rotation about the origin through π/2. (We drop reference to radian measure of angular displacement of the central angle of a unit circle since it is numerically equal to linear displacement on the circumference) Then we generate the coordinate axes by applying **j** on the unit vector **1** along the x-axis, *i.e.*, **j(1)** = **j**, the unit vector along the y-axis or **jy**, then on **j** to obtain **jj** = −**1**, the unit vector along the negative x-axis or −x, then on −**1**, to obtain **j(−1)** = −**j**, the unit vector along the negative **j**y-axis or −**j**y, and

then on −**j** to obtain −**jj** = **1**, back to the unit vector along the x-axis or x. The cyclic values of the composites of **j** are:

$$\mathbf{j}, \mathbf{j(j)} = \mathbf{j}^2 = -1, \mathbf{j(j^2)} = \mathbf{j}^3 = \mathbf{j(i)}, \mathbf{j(j^3)} = \mathbf{j}^4 = 1. \tag{105}$$

We define $-\mathbf{j}$ as inverse operator of $\mathbf{j}$, *i.e.*, clockwise rotation of the unit vector in the x-axis by $\pi/2$. Note: $-\mathbf{j(1)} = \mathbf{j(-1)}$. Applying composite mappings on the unit vector $\mathbf{1}$ along the x-axis successively, we have the four cyclic images of $\mathbf{1}$ in (105) under the composites of the operator $\mathbf{j}$ along the $\mathbf{j}$y-, $-$x-, $-\mathbf{j}$y and x-axes, respectively. For $n > 4$, the cycle is repeated and we define $\mathbf{j}^n = \mathbf{j(j^{n-1})}$, $n - 1, 2, \ldots,$ where we define $\mathbf{j}^0 = \mathbf{1}$.

Scalar and Vector Operations

For completeness, we introduce scalar multiplication. If α is an element of $\mathbf{R^*}$ [3], called scalar, $\alpha\mathbf{j} = \mathbf{j}\alpha$ is a vector of modulus a along the $\mathbf{j}$y-axis so that a commutes with $\mathbf{j}$. If b is another scalar,

$$(\alpha\beta)\mathbf{j} = \mathbf{j}(\alpha\beta) = (\mathbf{j}\alpha)\beta = \beta(\mathbf{j}\alpha) = (\beta\alpha)\mathbf{j}, \tag{106}$$

which follows from $\mathbf{j}$ being left-right operator and the commutativity of multiplication in $\mathbf{R^*}$. From commutativity and associativity of multiplication we have, for $\alpha, \beta, \gamma \in \mathbf{R^*}$,

$$(\alpha\beta\gamma)\mathbf{j} = \alpha(\beta\gamma)\mathbf{j} = (\alpha\beta)(\gamma\mathbf{j}) = (\beta\alpha)\gamma\mathbf{j} = (\gamma\beta\alpha)\mathbf{j} = \mathbf{j}(\gamma\beta\alpha). \tag{107}$$

Also, from distributivity of multiplication in R* with respect to addition we have,

$$\alpha(\beta + \gamma)\mathbf{j} = (\alpha\beta + \alpha\gamma)\mathbf{j} = (\alpha\beta)\mathbf{j} + (\alpha\gamma)\mathbf{j}. \tag{108}$$

Thus, we have retrieved the basic properties of the complex plane. We call $\mathbf{j}$ the complex vector operator.

Every vector in the complex vector plane has new real and complex components; conversely, a vector is the vector sum of its new real and complex components. Thus, a vector $\mathbf{z}$ in it has standard form,

$$\mathbf{z} = \alpha + \beta\mathbf{j} \text{ or } \mathbf{z} = (\alpha, \beta\mathbf{j}), \alpha, \beta \in \mathbf{R^*}. \tag{109}$$

The arithmetic of the complex plane holds provided that whenever $\mathbf{1}$ appears as a factor we interpret it as a unitary vector operator so that $\mathbf{1}\alpha = \alpha$, the vector of modulus a along the x-axis. Thus, we retain in the complex vector plane the vector algebra of the complex plane, the latter embedded in the former isomorphically. All concepts of the complex plane except i which is replaced by $\mathbf{j}$ carry over to the complex vector plane. For example, the norm or modulus of complex vector $\mathbf{v} = \alpha + \beta\mathbf{j}$, denoted by $|\mathbf{v}|$, is given by,

$$|\mathbf{v}| = (\alpha^2 + \beta^2)^{1/2}, \tag{110}$$

the square root of the product of $\mathbf{z}$ and its conjugate, $\alpha - \beta\mathbf{j}$. The dot product of vectors $\mathbf{u}$ and $\mathbf{v}$ is given by

$$\mathbf{u \cdot v} = |\mathbf{u}| |\mathbf{v}| \cos\theta \text{ if } \mathbf{u} \neq \mathbf{0}, \mathbf{v} \neq \mathbf{0}, \mathbf{u \cdot v} = 0 \text{ if } \mathbf{u} = \mathbf{0} \text{ or } \mathbf{v} = \mathbf{0}, \tag{111}$$

where θ is a new real number.

We use an arrow with initial and terminal points to represent a vector. Two parallel vectors with the same norm are equivalent. Therefore, a vector can be translated so that its initial point lies in the origin. This is called standard vector and has standard representation of the form (109).

The vector additive and multiplicative identities are $\mathbf{0}$ and $\mathbf{1}$, respectively, where the latter called the unit vector coincides with its real component 1. The scalars (new reals) are subject to the operations in $\mathbf{R^*}$ and scalar multiplication on vectors which are both commutative and associative [19]. With the complex vector arithmetic now defined we have verified that the operator $\mathbf{j}$ applies to any vector in the complex vector plane. Applying $\mathbf{j}$ on the vector $\mathbf{z}$ of (108), we have,

$$\mathbf{j(z)} = \mathbf{j}(\alpha + \mathbf{j}\beta) = -\beta + \mathbf{j}\alpha, \tag{112}$$

a positive rotation of vector **z** by $\pi/2$.

The Operator h_θ

We introduce a more general complex left-right operator on the complex vector plane appropriate for analytical work:

$$\mathbf{h}_\theta(a+\mathbf{j}b) = r(\cos\theta + \mathbf{j}\sin\theta), \tag{113}$$

a rotation of the unit vector **1** around the terminal point of the vector $\alpha + \mathbf{j}\beta$ by θ, an element of **R***. We can represent a vector **z** in the complex vector plane as,

$$\mathbf{z} = \mathbf{h}_\theta\alpha + \mathbf{j}\beta) = r(\cos\theta + \mathbf{j}\sin\theta), \tag{114}$$

where r is the modulus of **z** and θ its argument. This is the well defined counterpart of $e^{i\theta}$. If we vary α and β along **R*** and θ in $[0, 2\pi]$ the terminal point of **z** covers the entire complex vector plane. Geometrically, r varies in $[0, \infty)$ and rotates around the origin from 0 to 2π as the unit circle with center at the terminal point of $\alpha + \mathbf{j}\beta$ rotates from $\theta = 0$ to $\theta = 2\pi$. Then a point z_0 is given by

$$\mathbf{z}_0 = \mathbf{h}_\theta(\alpha_0 + \mathbf{j}\beta_0) = r_0(\cos\theta_0 + \mathbf{j}\sin\theta_0), \ \theta_0 \in [0, 2\pi]. \tag{115}$$

If $r_0 = 1$, (115) reduces to the equation of the unit circle with center at the origin. This operator applied on a vector along the x-axis rotates it by θ, reducing to operator **j** when $\theta = \pi/2$.

In the solution of the gravitational n-body problem [43], the operator that generates the spiral covering of a vortex is a variant of $\mathbf{h}_\theta$ and has the form,

$$z = ae^{\lambda t}\eta(t), \tag{116}$$

where η is given by the expression,

$$\mathbf{h}_\lambda(t) = \cos\lambda + \mathbf{j}e^{-t}\sin\lambda, \tag{117}$$

depending on the specific cases and phases of the evolving boundary conditions of the problem; here λ is constant of integration in the solution of the constraint equation of the associated optimal control formulation of this problem [117].

Suppose vector **z** has initial and terminal points $(\alpha, \mathbf{j}\beta)$ and $(\gamma, \mathbf{j}\zeta)$, respectively. Then,

$$\mathbf{z} = \mathbf{z}_1 - \mathbf{z}_2 = (\alpha, \mathbf{j}\beta) - (\gamma, \mathbf{j}\zeta).$$

Therefore,

$$\mathbf{j}(\mathbf{z}) = \mathbf{j}(\mathbf{z}_1 - \mathbf{z}_2) = \mathbf{j}((\alpha + \mathbf{j}\gamma\beta) - (\gamma + \mathbf{j}\zeta))$$

$$= \mathbf{j}(\alpha - \gamma) + (\zeta - \beta) = (\zeta - \beta) + \mathbf{j}(\alpha - \gamma), \tag{118}$$

is a counterclockwise rotation of **z** by $\pi/2$. In general, a polygon of n edges $e_1, ..., e_n$ may be represented as the vector sum $e_1 + ... + e_n$ or its resultant **r**. Then $\mathbf{j}(e_1 + ... + e_n) = \mathbf{j}(\mathbf{r})$ is a counterclockwise rotation of the polygon by $\pi/2$.

The operator **j** is an automorphism of the complex vector plane. Its additive inverse $\square$**j** is clockwise rotation about the origin by $\pi/2$.

There is, however, a new vector operation in the complex vector plane that may not have counterpart in other vector spaces: the product of two vectors. Let $\mathbf{u} = \alpha + \mathbf{j}\beta$, $\mathbf{v} = \gamma + \mathbf{j}\zeta$ the their product is given by

$$\mathbf{uv} = (\alpha + \mathbf{j}\beta)(\gamma + \mathbf{j}\zeta) = (\alpha\gamma - \beta\zeta) + \mathbf{j}(\alpha\zeta + \beta\gamma) \tag{119}$$

that, restricted to the complex vector plane, reduces to standard complex vector multiplication with $\mathbf{j}$ replaced by i. This is a particularity of the complex vector plane not shared by other vectors spaces. Consider vectors,

$$\mathbf{z}_1 = (\alpha + \mathbf{j}\beta),\ \mathbf{z}_2 = (\gamma + \mathbf{j}\zeta); \tag{120}$$

then,

$$\mathbf{z}_1\mathbf{z}_2 = \mathbf{h}_\theta(\alpha+\mathbf{j}\beta)\mathbf{h}_\phi(\gamma+\mathbf{j}\zeta) = \mathbf{r}_1\mathbf{r}_2\ (\cos(\theta+\phi) + \mathbf{j}\sin(\theta+\phi)), \tag{121}$$

where r_1, r_2 are the respective moduli of the vectors of $\mathbf{z}_1$, $\mathbf{z}_2$ and θ, ϕ their arguments. Note that this product of complex vectors is distinct from both the dot and vector products in a vector space. It is an extension of multiplication of complex numbers. Since the product of two complex vectors is a complex vector the product vector can be extended to any number of factors.

The additive inverse of a complex vector is quite obvious. For the multiplicative inverse we reduce its reciprocal to standard form. For instance, if $\mathbf{z} = \alpha + \mathbf{j}\beta$ then its multiplicative inverse $\mathbf{z}^{-1}$ is given by

$$\mathbf{z}^{-1} = 1/(\alpha+\mathbf{j}\beta) = (\alpha - \mathbf{j}\beta)/(a^2+b^2)$$

$$= \mathbf{h}(\mathbf{z}^{-1}) = \mathbf{j}(1/r)(\cos\theta - \mathbf{j}\sin\theta), \tag{122}$$

where $1/r = \mathrm{mod}\ (\mathbf{z}^{-1})$ and $\theta = \arg(\mathbf{z}^{-1})$. Then division of complex vector by another reduces to its multiplication by the inverse of the other. In general, if $\mathbf{z}_1 = r_1(\cos\theta + \mathbf{j}\sin\theta)$, $\mathbf{z}_2 = r_2(\cos\phi + \mathbf{j}\sin\phi)$, then $\mathbf{z}_1/\mathbf{z}_2 = (r_1/r_2)(\cos\theta - \sin\phi)$. Note that the operator $\mathbf{h}_\theta$ is really equivalent to the old notation $e^{\mathbf{j}\theta}$ and the latter may be used in place of $\mathbf{h}_\theta$ when convenient.

The operator $\mathbf{j}$ played a crucial role in solving the gravitational n-body problem [41] by generating the spiral covering of the underlying vortex by the gravitational flux streamlines as solutions of the conjugate equations obtained by the integrated Pontrjagin maximum principle from the optimal control formulation of this problem [117]. The n bodies and their rotating trajectories were obtained along specific spiral streamlines by the fractal-reverse-fractal algorithm [36] using a body at the core of the vortex as fractal generator. Elliptical orbit in the underlying spinning vortex is attained when the gravitational flux pressure balances the centrifugal force, its ellipticity being due to radial fluctuation of this balance by virtue of the oscillation universality principle, another expression of perfect balance being unstable which accounts for the fact that orbits of cosmological bodies are elliptical.

This short note provides the right setting for complex vector analysis on the complex vector plane as extension of **R***.

THE GENERALIZED INTEGRAL

The materials in this section are mainly quoted from [19]. We utilize wild oscillation to construct the generalized integral using approximation by rapid oscillation [19]. A key principle in the construction is a mathematical model of the complementarity of existence and speed in physics of which Heisenberg uncertainty principle [83] is special case:

Oscillation Probability Principle

The normalized derivative dq/dw of the approximating rapid oscillation $g(x) = \sin^2 x$ is the probability that the projection of the oscillating point P lies outside the subinterval [y, y+dy) in the set-value of the given wild oscillation.

This principle is an example of physical mathematics, *i.e.*, mathematics derived from physical principle. Physical mathematics has now become a tool in both physics and mathematics [49].

We shall use this principle to derive another key concept of generalized integration, probability distribution.

Problems that are impossible or difficult to solve by computation alone are often easily disposed of by qualitative analysis, *i.e.*, pure reasoning based on mathematical or physical principles.

Probability Distribution

A set without structure is uninteresting. For our purposes the appropriate structure on set-valued function is probability or unit measure distribution. Measure distribution is any distribution of entities, e.g. density and pressure. Then the sum of density or pressure over a distance, an area or volume is force. It can be a variable. For instance, the water pressure along a vertical line is a function of depth. When the weighted average of these entities is divided by their total sum we call the quotient probability or unit measure distribution, *i.e.*, normalized probability distribution. Since distribution is a sum we can integrate a function with respect to it and when the function varies over the range of a set-valued function such integral is called a generalized integral; it is particularly designed for integrating set-valued function with distribution, not necessarily probability distribution. For instance, it can be used for calculating the total force on a dam through generalized integration with respect to the pressure distribution. We shall apply the generalized integral to quantum gravity in the next chapter. However, since probability distribution is simply normalized distribution of any kind such as pressure and density it has broad applications. It should be particularly useful when the distribution is not homogeneous. For example, blood pressure on the body is not only variable but depends on a number of factors such as gravity, pumping of the heart and the position of the body (e.g., when lying down, sitting or standing) and blood vessels blockages due to a disease called stenosis.

The Wild Oscillation $Sin^m 1/x$ and its Extension to $Sin^m 1/0$

Consider the wild oscillation, $\{f(x)\} = \sin^m 1/x$ where n is a large integer. We use the oscillation probability principle to derive its expectation using the rapid oscillation $W(x) = \sin^m nx$. We consider the more general case m an even integer since when m is odd the computation is trivial in view of the symmetry of the probability distribution involved. The function $\{f(x)\}$ is set-valued only at $x = 0$, its set-value being the vertical interval segment $[0, 1]$ denoted by $\sin^m 1/0$, along the y-axis. We put structure on it, its cross-sectional probability distribution along the vertical segment $[0, 1]$ at $x = 0$. We approximate it by the probability distribution of the rapid oscillation $W(x)$ with m even and n a large integers. As $x \to 0$ the wild oscillation becomes more and more rapid and its arc on one period becomes more and more symmetrical with respect to the vertical through its maximum point and, therefore, is approximated more and more by the rapid oscillation $W(x)$.

We note further that multiplying a function by a constant only alters its values not its relative values and amounts to change of scale, the basis of normalization of distribution to turn it into probability distribution by dividing it by the sum of its values. Thus, for purposes of approximating probability distribution we do not even need the multiplier n; we use the ordinary function $f(x) = \sin^m x$ over the interval $[0, \pi/2]$ which corresponds to half a period of an arc above the x-axis since m is even. The effect of n is only to shrink its period so that the arc will approximate one period of an arc of the wild oscillation $\{f(x)\}$. Therefore, n is not necessary because it has no effect on the approximating probability distribution. From the symmetry of the rapid oscillation we need only a half arc over the interval $[0, \pi/2]$. Moreover, m only determines the bluntness at the maximum and flatness of base of the approximating arc and has insignificant effect on the distribution of values. Therefore, there is no loss of generality if we use the value $m = 2$.

Let x increase uniformly from 0 to $\pi/2$ then the projection of the point $P(x, w(x))$ on the y-axix sweeps over the set-value $[0, 1]$ of $\{f(x)\}$ at $(0, 0)$. Divide $[0, 1]$ by the non-overlapping subintervals $\{[y, y+dy)\}$, as y ranges from 0 to 1 except the interval on the upper end of the segment which we take as $[y, 1]$, $dy = 0$ (we assume this exception from now on). We calculate the probability distribution in terms of the variation or distribution of derivative, *i.e.*, relative variation of speed of the projection of P, over the half-arc in one sweep. Its speed on any of the subintervals is proportional to the derivative along corresponding subinterval of this half-arc. We drop the proportionality constant since that will be absorbed by the normalization of the probability distribution. We ask: what is the probability that the projection of P lies outside the subinterval $[y, y+dy)$? By the oscillation probability principle, it is proportional to the speed of the projection of P, *i.e.*, the derivative of dy/dw; again we drop the probability constant.

Denoting by dp/dw the probability that the projection of P on the interval lies in the subinterval [y, y+dy) and by dq/dw the probability that it lies outside the subinterval we have,

$$dp/dw + dq/dw = 1 \text{ or } dp/dw = 1 - dq/dw, \tag{123}$$

where w is a dummy variable for differentiation and, later, for integration. Since

$$dq/dw = 2\sin w\cos w, \tag{124}$$

we have,

$$dp/dw = (1 - 2\sin w\cos w)dw, \tag{125}$$

where dp/dw may not be normalized. To normalize (125) we first note that the projection of P lies in the interval [0, 1]; therefore, we divide dp/dw by the integral,

$$\int_{[0, \pi/2]} (1 - 2\sin w\cos w)dw = (w - \sin^2 w)|_{[0, \pi/2]}$$

$$= 1 - \pi/2 = (2 - \pi)/2. \tag{126}$$

We take the positive value of this normalizing constant; then the normalized probability distribution is given by

$$dp/dw = 2(2\sin\cos w - 1)/(\pi - 2)))dw. \tag{127}$$

Computing the expectation of the set value [0, 1] of {f(x)} and calling it generalized derivative (GD), we have,

$$GD(\{f(x)\}) = E(\{f(x)\})$$

$$= 2\int_{[0, \pi/2]} ((2\sin w\cos w - 1)/(\pi - 2)))\sin^2 w\, dw$$

$$= \int_{[0, \pi/2]} 2((2\sin^3 w\cos w - \sin^2 w)/(\pi - 2))dw$$

$$= 2((1/2)\sin^4 w) - (w/2 - (1/4)\sin 2w))/(\pi - 2))|_{[0, \pi/2]}$$

$$= 2((-1/2) + (\pi/4))/(\pi - 2)$$

$$= (2/4)(\pi - 2)/ (\pi - 2) = 1/2. \tag{128}$$

This is the approximate expectation of {f(x)} = $\sin^m 1/x$. The inflection point of the rapid oscillation at x = π/4, and f(π/4) = 1/2 distorts the actual distribution; it has the same effect on distribution as compact support of a function: its values are concentrated in it. In this case its counterpart is the singleton {1/2}. Normally, without the inflection point, the expectation of both the approximating half-arc of f(x) and E({f(x)}) would have been near the base of the curve on the x-axis due to the flatness of both f(x) and an arc of {f(x)} near the origin but the inflection point where the projection of P stops momentarily, skews the probability distribution up and makes it coincide with the inflection point, *i.e.*, GD({f(x)}) = 1/2. Therefore to avoid the distortion we calculate the sum of the approximated expectations in the subintervals [0, 1/2] and [1/2, 1], *i.e.*,

$$E(\{f(x)\})_{[0, 1]} = E_{[0, 1/2]}(\{f(x)\}) + E_{[1/2, 1]}(\{f(x)\})$$

$$= \int_{[0, \pi/4]} 2(2\sin\cos w - 1)/(\pi - 2)))dw$$

$$+ \int_{[\pi/4, 1]} 2(2\sin\cos w - 1)/(\pi - 2)))dw$$

$$= 2((1/2)\sin^4 w) - (w/2 - (1/4)\sin 2w))/(\pi - 2))|_{[0, \pi/4]}$$

$+ 2((1/2)\sin^4 w)$

$- (w/2 - (1/4)\sin 2w))/(\pi - 2))\big|_{[\pi/4,\,\pi/2]}$

$= 2(-1/8 + 1/4 + \pi/8) - 1/4 + \pi/4))/(\pi - 2))$

$= (3\pi/4 - 1/4)/(\pi - 2) = 0.3,$ **(129)**

the actual approximation of $E(\{f(0)\})$.

Since the probability distribution of $\{f[x]\} = \sin^m 1(x - s)$, as $x \to s^+$, $s \in [0, \pi]$ is uniform it is constant in the interval $[0, \pi]$, *i.e.*, $y = E(\{f[x]\}) = 0.3$. Therefore, the approximate weighted area of the extended wild oscillation $\{f[x]\}$ in the interval $[0, \pi]$ is given by

$\int_{[0,\,\pi]}\int_{[0,\,\pi/4]} 2\sin^m w\,((2\sin w\cos w - 1))/\,(\pi - 2))dwdx = \int_{[0,\,\pi]}(1/2)dx = 0.3\pi.$ **(130)**

Consider the integral,

$F(x) = \int_{[0,\,x]}(2\int_{[\pi/4,\,\pi]}(2\sin\cos w - 1)\sin^2 w/(\pi - 2))dwdx + 2\int_{[0,\,\pi/4]}\int_{[\pi/4,\,\pi/2]}(\sin\cos w - 1)\sin^2 w/(\pi - 2))dwdx.$ **(131)**

Then we define the derivative of the double integral as the inner integral and we have an analogue of the fundamental theorem of the calculus:

$(d/dx)F(x) = (2\int_{[\pi/4,\,\pi]}(2\sin\cos w - 1)\sin^2 w/(\pi - 2))dw + 2\int_{[\pi/4,\,\pi/2]}(\sin\cos w - 1)\sin^2 w/(\pi - 2))dw.$ **(132)**

The Wild Oscillations $(\sin^m 1/x)(\cos^m 1/x)$, $(\sin^m 1/0_x)(\cos^m 1/0_x)$

We derive the scheme for finding the probability distribution of the product wild oscillation (c) given by $\{g(x)\} = (\sin^m 1/x)(\cos^m 1/x)$, its set value being $(\sin^m 1/0)(\cos^m 1/0)$ at $x = 0$. We approximate the derivative by the appropriate rapid oscillation $g(x) =$

$\sin^m n\pi x)(\cos^m n\pi x$. Again, without loss of generality, we let $m = 2$, $n = 1$ so that

$dq/dw = (d/dw)(\sin^2 w\cos^2 w)$

$= 2(\sin w\, x\cos^3 w - \cos w\sin^3 w)dw.$ **(133)**

This product function is symmetric with respect to the vertical line at its maximum. To get the maximum, let

$\mathrm{Sin}\,x\cos x(\cos^2 x - \sin^2 x) = 0.$ **(134)**

Then, the solutions in the interval $[0, \pi/2]$ are $x = 0$, $x = \pi/2$, $x = \pi/4$ so that the maximum is at the midpoint $x = \pi/4$ and the minima are at the two end points $x = 0$ and $x = \pi/2$.

Just to have a sense of how the product function looks like we note that $f(x)$ increases from 0 to 1 in the interval $[0, \pi/2]$ and $g(x)$ decreases from 1 to 0 in the same interval. They intersect at $x = \pi/4$, $f(\pi/4) = (\sqrt{2})/2$. To the left of the intersection, $f(x) < g(x) < 1$ and to the right $g(x) < f(x)$ and at the intersection $f(x) = g(x)$. It follows that, $0 \le f(x)g(x) < g(x)$ on the left and $0 \le f(x)g(x) < f(x)$ on the right. Therefore, the curve lies under both curves and its maximum is beneath their intersection so that it has no inflection point (see graph in [10]). Since $f(x)$ and $g(x)$ intersect at their inflection point their product has no inflection point. Moreover, since the product function has degree 4 in the sine and cosine functions which are both less than 1 its expectation must be very small, very close to the x-axis. Note that we obtain more information about the problem by qualitative analysis. (Incidentally, the product of an oscillation with any function is an oscillation)

Since, the sine and cosine functions are half a period shift from each other the product function is symmetric with respect to the vertical through its maximum. We, again, find the probability distribution on a half arc, i.e, the image of the interval $[0, \pi/4]$ on the set value $[0, (\sqrt{2}/2)]$ along the y-axis under the product function.

We subdivide the vertical interval $[0, (\sqrt{2}/2)]$ at the origin by the non-overlapping subintervals $\{[y, y+dy)\}$. Since this product function is symmetric the two branches of the arc have the same probability distribution and expectation so that it suffices to find the probability distribution in the interval $[0, \pi/4]$.

We apply the oscillation probability principle again on the vertical interval $[0, (\sqrt{2}/2)]$. Since

$$dq/dw = 2(\sin w \ x\cos^3 w - \cos w \sin^3 w)dw, \tag{135}$$

then by the oscillation probability principle we have,

$$dp/dw = 1 - 2(\sin w \ x\cos^3 w - \cos w \sin^3 w)dw. \tag{136}$$

We find the normalizing constant,

$$\int_{[0, \pi/4]} (1 - 2(\sin w \ x\cos^3 w - \cos w \sin^3 w)dw$$

$$= -\pi/4 - 2\int_{[0, \pi/4]}(\sin w \ x\cos^3 w)dw + 2\int_{[0, \pi/4]}\cos w \sin^3 w)dw$$

$$= -\pi/4 - 2(-1/4)\cos^4 w|_{[0, \pi/4]} + 2(1/4)\sin^4 w|_{[0, \pi/4]}$$

$$= -\pi/4 + 1/2 \ \cos^4 w|_{[0, \pi/4]} + 1/2 \ \sin^4 w|_{[0, \pi/4]}$$

$$= -\pi/4 + 2/8 = (1 - \pi)/4. \tag{137}$$

We take the positive value, $(\pi - 1)/4$ for the normalizing constant and reverse the terms of dq/dw to make it positive in the interval. Therefore, the normalized probability distribution is,

$$dp/dw = 4(2(\sin w \ x\cos^3 w - \cos w \sin^3 w) - 1)/(\pi - 1))dw. \tag{138}$$

We evaluate the approximate expectation or generalized derivative.

$$E(\{f(x)g(x)\}) = 4\int_{[0, \pi/4]} (2(\sin w\cos^3 w - \cos w \sin^3 w)$$

$$- 1)(\sin^2 w\cos^2 w)/(\pi - 2))dw$$

$$= 4\int_{[0, \pi/4]} (2(\sin^3 w\cos^5 w - \sin^5 w\cos^3 w)$$

$$- \sin^2 w\cos^2 w)/(\pi - 1))dw$$

$$= 4\int_{[0, \pi/4]} (2(\sin^3 w(1 - \sin^2 w)^2\cos w$$

$$- \sin^5 w(1 - \sin^2 w)\cos w) /(\pi - 1))dw$$

$$+ 4(\sin^2 w(\sin^2 w - 1)/(\pi - 1))dw$$

$$= 4\int_{[0, \pi/4]} (2(\sin^3 w(1 - 2\sin^2 w + \sin^4 w)\cos w$$

$$- (\sin^5 w - \sin^7 w)\cos w/(\pi - 1))dw$$

$$+ 4\int_{[0, \pi/4]} (\sin^4 w - \sin^2 w))/(\pi - 1))dw$$

$$= 4\int_{[0, \pi/4]} 2(\sin^3 w - 2\sin^5 w + \sin^7 w)\cos w$$

$$- (\sin^5 w - \sin^7 w)\cos w/(\pi - 1))dw$$

$$+ 4\int_{[0, \pi/4]} (\sin^4 w - \sin^2 w)/(\pi - 1))dw$$

$$= 4((1/2)\sin^4 w - 2(1/6)\sin^6 w + (1/8)\sin^8 w$$

$$- (1/6)\sin^6 w/(\pi - 1))|_{[0, \pi/4]}$$

$$+ 4((1/8)\sin^8 w)/(\pi - 1))|_{[0, \pi/4]} + 4((-(w/2$$

$$+ (1/4)\sin 2w)|_{[0, \pi/4]} - \sin^3 w \cos w/(\pi - 1))|_{[0, \pi/4}$$

$$+ 4(1/3)\int_{[0, \pi/4]} \sin^2 w)/(\pi - 1)dw$$

$$= 4((-(1/8) + 1/24 - 1/128 + 1/48$$

$$- (1/28)/(\pi - 1) + (\pi/8 - (1/4)/(\pi - 1)$$

$$+ 4(1/4)(1/4) + (1/3)(w/2 \ (1/4)\sin 2w/(\pi - 1))|_{[0, \pi/4]} = 0.07. \tag{139}$$

Thus, the expectation or weighted average is very close to the base.

The probability distributions and expectations of the functions,

$$y = \sin^n(1/x^2)\cos^m(1/x^2) \text{ and } y = \sin^2(1/x^2) + \cos^m(1/x^2) \tag{140}$$

are derived in [2] from which L'Hospital's rule is generalized for these functions. L'Hospital's rule does not apply to them because they have zeros in every neighborhood of the origin.

CHAPTER 3

The Grand Unified Theory

Abstract: We provide a detailed account of the successful 5,000-year-old search for the superstring, basic constituent of matter, capped by its capture in 1997, and highlights what particle physics have accomplished as a vital link in the capture of the superstring. The latter was the key to the solution of the unsolved problems and resolution of fundamental questions of physics, particularly, the 200-year-old gravitational n-body and turbulence problems and about the structures of the superstring, electron, proton, neutron and atom. Most of all, it laid the foundations of the grand unified theory that binds the forces and interactions of nature in this single theory. Naturally, it advances and unifies the natural sciences and strengthens multidisciplinary research as well. As a matter of fact, its theoretical applications have opened up new fields in biology, geology and oceanography, atmospheric science, physical psychology, astronomy, cosmology and medicine.

GUT explains some previously puzzling phenomena, e.g., superconductivity, matter-anti-matter interaction and metal fatigue, and corrects misinterpretation of natural phenomena, e.g., thermonuclear reaction, magnetic levitation and supernova. Most of all, it brings us to the threshold of a new technological epoch that utilizes the clean, inexhaustible and free dark matter, one of the two fundamental states of matter, that is abundant everywhere in the Cosmos.

The crucial element for all these discoveries are qualitative mathematics and modeling; the latter explains not only the appearances of nature but also how nature works in terms of its laws. Qualitative mathematics is the only route towards the discovery of natural laws.

INTRODUCTION

This chapter requires much imagination as it deals with physical systems, *i.e.*, motion or configurations of matter, that are not directly observable such as dark matter and its basic constituent, the superstring. Moreover, the subject matter can be understood only from the standpoint of the grand unified theory (GUT) not conventional physics that can only describe natural phenomena. GUT alone as physical theory explains how nature works through the discovery and applications of natural laws and rational thought and analysis.

Key to the development of GUT was the discovery of the superstring in 1997 [41]. During the last half century physicists have been smashing the nucleus of the atom in search of the irreducible elementary particles. By the 1990s they had completely succeeded with the discovery of the +quark (up quark) and –quark (down quark) that, together with the electron (discovered in 1897 by J. J. Thompson), constitute the irreducible elementary particles or basic prima that comprise the light isotope of every atom. The heavy isotope of an atom has additional constituent, the neutrino in the neutron [43]. The conversion of the superstring to the basic and non-basic prima, units of visible matter, is the vital link in establishing it as the basic constituent of matter. Among the non-basic prima are the neutrino and the anti-matter of the basic prima. Like the electron, the superstring is unique and replicated everywhere in the Cosmos in three forms.

The basic or non-agitated and semi-agitated superstrings that comprise dark matter could not have been found in the nucleus, nor could it have been discovered by quantitative modeling that relies on observation, measurement and computation. It needed a new methodology, qualitative modeling, applied to physics for the first time to solve the 200-year-old gravitational n-body problem in 1997 [41] that relies on qualitative mathematics introduced for the first time in and the main contribution of [46]. Being dark, the non-agitated and semi-agitated superstrings are not directly observable and known only by their impact on visible matter and explanation provided by qualitative modeling. Moreover, all that have been found in the nucleus aside from the basic prima are unstable prima with half-life of split second and do not qualify as basic constituent of matter.

The superstring must be both stable and indestructible otherwise our universe would have collapsed a long time ago. Given the requirements on the superstring the only route to its discovery is theoretical and in the 1950s Paul Dirac succeeded in developing the first string theory that *describes* particles as loops of strings vibrating in higher dimensions as much as 26. However, strings are not directly observable and since they are supposed to have zero mass, do not occupy space and yet exert a force, there is clear violation of energy conservation. Although several string theories were developed since Dirac, no significant improvement over his initial work has been made and the quest continues. The ambiguity of the concept *dimension*, contributes to this difficulty aside from the ambiguity of *superstring*.

To be fair, dark matter consisting of semi- and non-agitated superstrings was unknown in the 1950s when particle physicists embarked on smashing the nucleus of the atom in search of the basic or irreducible elementary particles; by 1990 they had found them with astounding success but without fanfare.

We apply the full power of qualitative mathematics and modeling on the pillars of GUT – quantum and macro gravity and thermodynamics in the broad sense, *i.e.*, generation, conversion, transfer or conduction and utilization of energy. Utilization includes devising of technologies that run on dark matter called GUT technologies. We start a new direction in the search for the superstring that relies on the discovery of the appropriate laws of nature away from the traditional search for it in the atomic nucleus.

QUANTUM GRAVITY

The domain of quantum gravity includes dark matter so that, in a sense, it is an extension of quantum physics to dark matter. Its core of study is the atom that has all its dynamics and interactions.

The Last Stretch in the Search for Basic Constituent

So far, the superstring is just a name but we shall establish not only its existence but also its structure, properties or behavior and interactions using qualitative mathematics alone together with known facts and information about the Cosmos. We convert *indestructibility* into an operative or useful concept to pin down the structure and behavior of the superstring. Consider this thought experiment.

Imagine an egg shell that contains an egg shell that contains an egg shell, etc., ad infinitum. Such structure is called generalized nested fractal *sequence* (the terms are infinite). If we hit it with a hammer, can we destroy all the shells and their structure? We cannot. The hammer will hit the first shell and possibly a few more but the tail sequence will survive as a generalized nested fractal sequence of shells. This is the structure of the superstring we are looking for.

We start with what we consider the most fundamental law of nature and proceed to find others consistent with it. In the event that we encounter a natural phenomenon that appears to contradict it, we find another that reconciles them. This is assured by our premise that there is order in our universe. We take as the most fundamental law the first law of thermodynamics formulated by the German physician and physicist Julius Robert von Mayer in 1842 [12] (also associated with Rudolf Clausius who started axiomatization of thermodynamics [73]). It states that energy (matter) cannot be created or destroyed. We at once see its inadequacy: it does not take into account latent or dark energy. For example, when gasoline is ignited were does the burst of energy come from? Consider another thought experiment: Shoot a beam of light into a *vacuum* and turn it off. The photons will vanish without trace since there are no molecules to absorb the energy of the light beam and convert it to motion or kinetic energy in terms of rise in temperature of the vacuum. In view of our definition of energy as motion of matter, we reformulate and enrich and embellish this law with more components and call it by a new name.

Energy Conservation

In any physical system and its interaction, the sum of kinetic and latent energy is constant, gain of energy is maximal and loss of energy is minimal.

Observation through the space telescope Hubble reveals that matter forms steadily in the Cosmos initially as cosmic dust that gets drawn into and entangled in cosmological vortices, collects around their vortex eyes by suction and forms stars at the rate of one star per minute [3, 96, 102]. There are regions in the Cosmos called star nests that release stars more rapidly than others [96, 102]. It is also known that a grain of cosmic dust is oblong revealing its formation around the eye of a micro vortex and that it has crust and mantle like Earth. Our conclusion: By energy conservation, what appears to be a void or empty region of space is filled with tiny physical systems so small the size of each has order of magnitude less than that of the finest wavelength of light, our medium for observation, which is 10^{-14} meters. That region is called dark matter because its constituents are not detected by our medium for observation – light. Based on this information and observation we have the next law of nature.

Existence of Two Fundamental States of Matter

There exist two fundamental states of matter, dark and visible (ordinary); the former is indirectly observable but known only by its impact on visible matter.

Now, we ask this valid and non-vacuous question: what does dark matter consist off? Whatever that is we call it superstring, a mere name at this point that we shall embellish with structure, properties and behavior using appropriate laws of nature. From the indestructibility and non-reducibility of the basic constituent of matter, every physical system must be reducible to it. Therefore, the superstring is unique, *i.e.*, it has the same structure and properties as other superstrings and can differ only in form or phase and location. The next law says that energy conservation has many forms.

Energy Conservation Equivalence

Energy conservation has many expressions or forms: order, symmetry, economy, least action, optimality, efficiency, stability, self-similarity (nested fractal), coherence, resonance, quantization, synchronization, smoothness, uniformity, motion-symmetry balance, non-redundancy, non-extravagance, evolution to infinitesimal configuration, helical and related configuration, circular, helical, spiral and sinusoidal and, in biology, genetic encoding of characteristics, reproduction and order in diversity and complexity of functions, configuration and capability.

We refer to each component of this law physical principle. This law has many applications and facilitates identification of physical systems and determination of their structure and properties. For example, in 2003 physicists were excited by the discovery of the third quark in the nucleus of the atom outside the protons. They thought this was the *gluon* that supposedly binds the protons in the narrow confines of the nucleus despite the enormous repulsion between them. The third quark joins two positive quarks, one from each of two protons. However, by the principles of non-redundancy and non-extravagance as component of the energy conservation equivalence law that says nature does not create a physical system whose function is already being done by an existing physical system it follows that the third quark is really the negative quark that joins the two positive quarks of the proton and the excitement fizzled out since the latter was discovered in the fifties at FermiLab, Bata*v*ia, Illinois near Chicago. Thus, the gluon they were looking for had already been found. The profile of a primum also follows from this law.

Each of the configurations that express energy conservation is a universal configuration of matter. Thus, the sinusoidal profile of a wave (optimal balance between symmetry and motion: motion breaks symmetry and symmetry precludes motion) is seen in the motion of ocean waves and clock pendulum and nested fractal configuration in the roots and branches of a tree and the veins of its leaf that provides optimal efficiency in the absorption and distribution of nutrients from the ground. Another instance of the universality of fractal is nested fractal vortex of a galaxy that contains the vortices of its stars, each star vortex contains the vortices of its planets, each planetary vortex contains the vortices of its moons, etc., [25]. We shall examine in detail later how cosmological vortices form.

We next state a law of nature that is central to the dynamics of all natural vortices.

Flux-Low-Pressure Complementarity

Low pressure sucks matter and the initial chaotic rush of dark matter towards a region of low pressure stabilizes into local or global coherent flux; conversely, coherent flux induces low pressure around it.

This law is self-explanatory and was inspired by a high school experiment many years ago that goes this way: place two books of the same thickness flat on the desk back to back about three inches apart; place a soft tissue over the gap between them; blow under the tissue. Will the tissue fly off? No, it will be pulled and carried by the flux of air blown underneath. The flux creates low pressure around it and pulls the tissue away. It has applications to all vortices, dark or visible. For example, with considerable wind movement the barometer plunges down and probably portends an impending typhoon. The eye of a tornado sucks as a region of low pressure created by the vortex flux of gas molecules around it. Most of all, it explains why tornado and hurricane form and, naturally, provides the basis for controlling them, full for the former and partial for the latter.

Resonance

Maximum resonance between waves, oscillation and vibration occurs when they have exactly the same characteristics but arc length or its reciprocal, frequency, is the principal factor for resonance. The degree of resonance declines drastically with the difference in orders of magnitude of frequency. However, negligible resonance between waves that differ by orders of magnitude add up to a significant level at critically high order of magnitude of frequencies.

This law is central to interaction of physical systems discovered in analyzing the disastrous final flight of the Columbia Space Shuttle.

We identify the structure of the superstring. One physical system that interacts with the superstring is basic cosmic wave. Since the superstring is generated by the normal vibration of the atomic nucleus and propagated in all directions across dark matter and the nucleus is comprised of superstrings the wave length of its vibration has comparable order of magnitude to that of the superstring. Therefore, there is resonance. When basic cosmic wave hits a non-agitated superstring the following may occur: (a) it is thrown into dark matter and bounces erratically with superstrings until the energy imparted on it is dissipated by the collision and grinds to a halt or (b) it gets near its earlier path, gets sucked by it, by flux-low-pressure complementarity, and forms a loop with the original superstring travelling through it. By energy conservation and energy conservation equivalence, this path evolves into the optimal energy conserving configuration, namely, helical spiral loop which has maximal symmetry and smoothness. This is the basic configuration of the superstring described in Chapter 2. By the quantization principle of energy conservation equivalence this is a semi-agitated superstring, its toroidal flux a non-agitated superstring traveling through its cycle at enormous speed 7×10^{22} cm/sec. By the fractal principle of energy conservation equivalence, the superstring toroidal flux contains a superstring toroidal flux, etc., ad infinitum.

What is the dimension of the motion of the superstring that concerned Dirac? Since, as we shall see later, dimensions are directly related to identifiable *independent* directions of motion (degrees of freedom) and its motion is due to the impact of cosmic waves coming from all directions it must be huge. In general the relevant dimensions depend on the subject matter of study. Thus, in meteorology, the independent variables include temperature, humidity and visibility.

With this new information we are able to see a new possibility and still gain more information. That is the nature of qualitative mathematics. Possibilities (a) and (b) may occur only when the superstring concerned is non-agitated. Then another possibility comes to view: (c) the first term of the nested fractal superstring in (a) bulges to retain the speed of the toroidal flux despite the energy imparted by the basic cosmic wave that hits it, by energy conservation, turning it into a semi-agitated superstring its toroidal flux being non-agitated. We state this result as a law of nature.

Existence of Basic Constituent of Matter and its Generalized Nested Fractal Structure

The basic constituent of dark matter is the superstring. It is a helical loop and nested fractal sequence of superstrings or toroidal fluxes, with itself as first term; each toroidal flux in the sequence is a superstring having toroidal flux, a superstring, traveling at speed beyond that of light along its cycles, etc.; each superstring except the first, is contained in and self-similar to the preceding term in structure, behavior and properties.

That the Universe has no beginning follows from the nested fractal *sequence* of the toroidal flux of the superstring; that it has no end follows from its indestructibility; that it is boundless follows from flux-low-pressure complementarity since if there were a boundary so that the other side is empty, the superstrings from our side will rush to it and demolish the boundary.

Thus, the superstrings fill up dark matter in the timeless and unbounded Universe. A superstring is dark if its cycle length is less than 10^{-14} meters, non-agitated if $CL < 10^{-16}$ meters, semi-agitated if $10^{16} < CL < 10^{-14}$ meters. Then we introduce a third phase: agitated or visible, *i.e.*, a segment has $CL > 10^{-14}$ meters. Left alone without agitation a superstring shrinks steadily for it shortens the helical paths of toroidal flux, by energy conservation, its tail end becoming infinitesimal physical continuum [21, 25]. The boundaries between agitated, semi-agitated and non-agitated superstrings will be defined by experimental and theoretical physicists but any changes will affect neither GUT's validity nor its explanation of how nature works.

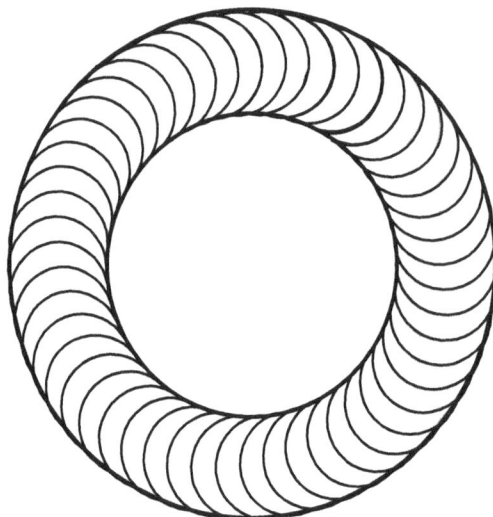

Figure 1: An artist's conception of a semi-agitated superstring. A non-agitated superstring, its toroidal flux, has a similar nested fractal structure as itself and travels through its helical cycles at the speed of 7×10^{22} cm/sec [4]. The helix winds around the torus rapidly its cycles infinitesimally close.

The Primum and Photon

What happens when a semi-agitated superstring is hit by basic cosmic wave? We state the answer as a law of nature.

Dark-to-Visible-Matter Conversion

When suitable shock wave hits a semi-agitated superstring one of these occurs: (a) the outer superstring breaks, its toroidal flux remaining non-agitated; (b) a segment bulges into a primum, unit of visible matter.

Energy conservation holds at all times in the cycle of a superstring including transition from one phase to another. The extreme cycle of the bulge in (b) is called the equator. While dark matter is inert, *i.e.*, dark superstring does not interact with anything else since it has infinitesimal induced flux, this is not the case with the primum. Its toroidal flux is hit from all directions by cosmic waves causing erratic motion called *spike* along the helical cycles as it speeds through it at 7×10^{22} cm/sec [4]. It pulls the superstrings around the primum and forms a vortex flux with its eye along the axis turning it into a magnet with polarity in accordance with the right hand rule of electromagnetism, i.e, when the index finger points in the direction of the toroidal flux, the thumb points to the N- or north pole; otherwise, it points to the opposite pole, the S- or south pole. The vortex flux is its magnetic flux whose energy is measured as charge. Thus, charge is energy or motion of matter, the primal induced vortex flux. A primum is positive if its vortex flux spins counterclockwise viewed from its north pole, negative otherwise. The electron is a negative primum, its charge −1 (1.6×10^{-19} coulombs [12]) and the unit of charge by convention. It is a basic primum. The positive quark, another basic primum, has charge +2/3 while the negative quark, the third basic primum, has charge −1/3 [52]. They are basic since they comprise every light isotope of an atom.

Coulomb's inverse square law that governs primal interaction is a description of the impact of charge on matter under its influence established empirically. It is analogous to Newton's law of gravitation which attests to the duality and similarity between electromagnetism and gravitation. Whether Coulomb's inverse square law can be derived from the laws of nature is an open problem.

The speed of the toroidal flux of a primum was deduced from the linear speed of the toroidal and induced vortex flux of the proton measured at 7×10^{22} cm/sec [4]. Since the toroidal flux is a superstring it resonates with and pulls the superstrings around the primum. Therefore, they have the same linear speed. Moreover, by the principle of synchronization of the energy conservation equivalence law, this linear speed must be true of all toroidal fluxes and their induced fluxes. Otherwise, there will be much dissipation of energy, a violation of energy conservation. Therefore, the speed of 7×10^{22} cm/sec is a constant of nature and applies to all toroidal and induced fluxes including electric current minus the effect of the conductor's resistance. Electric current is split off or by-pass from atomic vortex fluxes. Electrical power generation works on this principle. When a close circuit electrical conductor cuts across a magnetic line of force its vortex flux splits from the atomic nuclei and goes through the conductor (since it offers much less resistance than air) and becomes electric current.

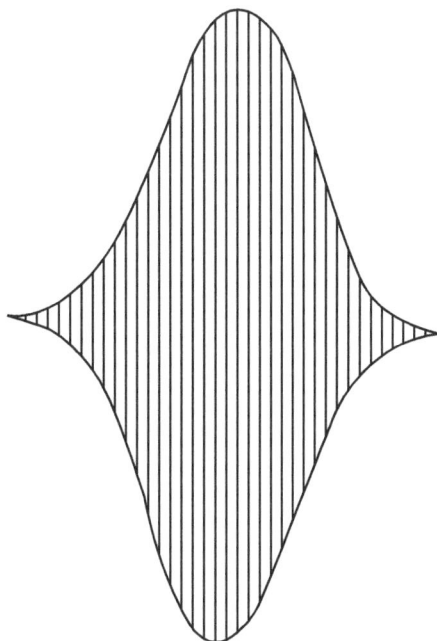

Figure 2: A simple primum, bulged segment of semi-agitated superstring and a magnet with induced vortex flux hidden. Its polarity conforms to the right-hand-rule of electromagnetism: when the index finger points to the direction of its toroidal flux, the thumb points to the N-pole.

Moreover, the frequency of the cycles of a primum can, theoretically, be computed from the energy of the photon it converts to when scooped up by basic cosmic wave that fits it by dividing by h. However, the ambiguity of small number [21] such as h makes this division similarly ambiguous. The best we can do is find the cycle or energy density of a primum or photon it converts to which requires the generalized integral [19]. A simple primum is charged so that when a primum is neutral it must be coupled. Thus, the neutrino must be coupled and by the principles of optimality, non-redundancy and non-extravagance it consists of simple prima of opposite but equal charges, say, $+q$ and $-q$ so that its charge is $+q + -q = 0$, *i.e.*, neutral. Charge calculation belongs to quantum algebra.

In cylindrical coordinates a simple primum has the equation $x = t$, $y(t) = \beta(\sin n\pi t)(\cos^m k\pi t)$, $\theta = n\pi t$, $t \in [-1/k, 1/k]$, n, m, k, integers, $n \gg k$, m even (n much larger than k) which means that the first factor of $y(t)$ consists of rapid oscillations and since this is in cylindrical coordinates, $y(t)$ consists of rapid spirals about the x-axis. The cycle energy of the spiral is Planck's constant $h = 6.64 \times 10^{-34}$ Joules [12]. Scooped up and carried by cosmic wave, its cycles flatten to rapid oscillations, $z = 0$, $x = t$, $y(t) = \beta(\sin n\pi t)(\cos^m k\pi t)$ due to dark viscosity. It becomes photon, $z = 0$, $y(t) = \beta(\sin n\pi t)(\cos^m k\pi t)$, when it breaks off from its loop; the energy of one full cycle of the primum or one full arc of the photon is h; its toroidal flux speed: 7×10^{22} cm/sec [4]. The configuration of the primum and photon is based on the universality of oscillation and related motion of matter and the uniformity principle of the energy conservation equivalence. (It follows that photons of different colors come from different prima).

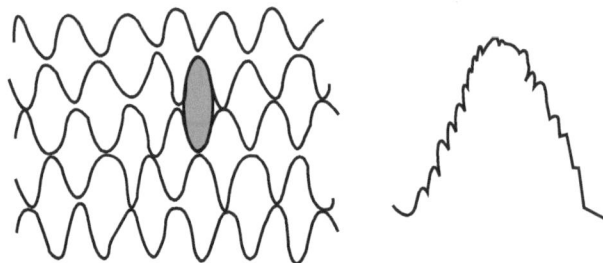

Figure 3: A segment of a beam of basic cosmic waves of the same order of magnitude; a primum or photon in flight (shaded) is lodged between two basic cosmic waves of opposite crests with wave length of the same order of magnitude as its envelope (left). An arc of a basic cosmic wave showing the first two terms of its nested fractal sequence (right).

A primum has angular momentum provided by the toroidal flux that travels through its helical cycles. However, when scooped up by suitable basic cosmic wave and becomes a rapid oscillation (planar, polarized light) due to dark viscosity its angular momentum converts to linear momentum which is augmented by that of the carrier; so does the photon. However, experiments show that both the primum in flight and the photon have angular momentum. Where does it come from? Neither the primum in flight nor the photon is free from the impact of vortex fluxes of superstrings that provide them angular momentum, not even in a vacuum since the latter is embedded in the Earth's gravitational or magnetic flux. This is even more so outside of vacuum since every atom has vortex flux. In fact, there is no place in the Cosmos free from the impact of some gravitational flux. In regions between galaxies, for example, gravitational flux is provided by the super…super galaxy, our universe [25].

Moreover, in any energy exchange, linear momentum does not necessarily convert to linear momentum nor does angular momentum to angular momentum. For example, the wind's linear momentum converts to angular momentum that rotates the sensor of the anemometer or the weather vane. Conversely, the angular momentum of the vortex flux of superstrings around a magnetic line of force of a magnet converts to the linear momentum of the electric current when the conductor cuts across it in an electric generator.

The relevant motion of matter in the primum is the toroidal flux through its cycles that endows it with. Therefore, the smallest unit of energy is the motion of the toroidal flux through one cycle and that is what we denote as the Planck's constant h. This is independent of the cycle length, by the principle of uniformity. In conventional physics, the energy of the photon in flight is given by $E = h\nu$, where ν is misinterpreted as the frequency of the carrier basic cosmic wave or the reciprocal of the arc length of its envelope.

For very long wavelengths and very low frequencies less than 1 Hertz, quantized energies less than h in magnitude are obtained (information provided by Prof. C. G. Jesudason, University of Malaya, in private correspondence). This contradicts the fact that h is the irreducible unit of energy. How do we account for it? The equation is valid only when ν is the frequency of the photon as rapid oscillation. Therefore, the equation does not apply to long wavelength radiation which does not carry a photon rider, by the resonance law, since the arc length of the envelope of the photon has different order of magnitude from the long wavelength of such radiation. Where does the energy of long wave radiation come from? It comes from the vibration of the generating source (an atomic nucleus in the case of pure electromagnetic wave) and the synchronized vibration of the medium, dark matter, which reinforces it. Therefore, the energy equation of such radiation must have terms that reflect the energy of the generating source and vibration of dark matter that resonates with and reinforces it. It must be a function of wavelength since long wavelength means less vigorous vibration and low energy of both source and medium. That equation which is unknown (open problem) at this time should apply to non-photonic (no photon rider) waves, e.g., water wave.

Although the energy of the photon is known, we cannot divide it by 6.64×10^{-34} to find the number of oscillations or cycles of the primum it comes from in view of the ambiguity of large and small numbers (dividing a number by the small number 6.64×10^{-34} is extremely inaccurate; this is a mathematical, not physical limitation). Therefore, we compute instead the energy density distribution of the rapid oscillation within the photon, *i.e.*, within its envelope given by $y(t) = \cos^m k\pi t$ in one period, $t \in [-1/2\pi k, 1/2\pi k]$ since m is even. Moreover, since actual accurate computation is impossible we need a series of approximations to find the energy density of the photon.

Consider the standard rapid oscillation,

$$y(x) = (\sin n\pi x), \quad x \in [0, 1/\pi k], \tag{1}$$

where is n large. It is approximated by the topologist sine curve with respect to variation of derivative, symmetry and other properties [71].

$$\{y(x)\} = \{\sin 1/(x)\}, \quad x \in \text{as } [0, 1/2\pi k], \ x \to 0^+, \tag{2}$$

at the origin where it is set-valued the interval $[-1, 1]$ its set-value (see Fig. **4a**). Therefore, the extended topologist sine curve,

$$\{y(x)\} = \text{Slim}(\{\sin 1/(x-s)\}), \quad x \in [0, 1/2\pi k], \text{ as } s \to x^+, \tag{3}$$

approximates the rapid oscillation (1) in the interval $[0, 1/2\pi k]$ for large n, where $\text{Slim}(\{\sin 1/(x-s)\})$ is the set limit of $\{\sin 1/(x-s\}$ called rectangular function (see Fig. **4**), *i.e.*, each full arc of the extended topologist sine curve approximates the variation of derivative and, therefore, the shape of a full oscillatory arc at each $x \in [0, 1/2\pi k]$ in a suitably small neighbourhood. Conversely, the latter approximates the former in the same sense. Note that in both arcs the maximum is 1 and the minimum -1 (a full arc of the rapid oscillation is symmetric with respect to its midpoint in the x-axis).

We recall that the product of an oscillation is an oscillation. The equation of the photon at the origin that travels with the xy-plane in the direction of the x-axis is the rapid oscillation given by,

$$y(x) = \beta(\sin n\pi x)(\cos^m k\pi x), \quad x \in [0, 1/2\pi k], \tag{4}$$

is also a rapid oscillation whose envelope is the curve

$$y(x) = \pm\beta(\cos^m k\pi x), \quad x \in [0, 1/2\pi k]. \tag{5}$$

We note that the product of a rectangular function with an oscillation is a set-valued with the latter as the envelope.

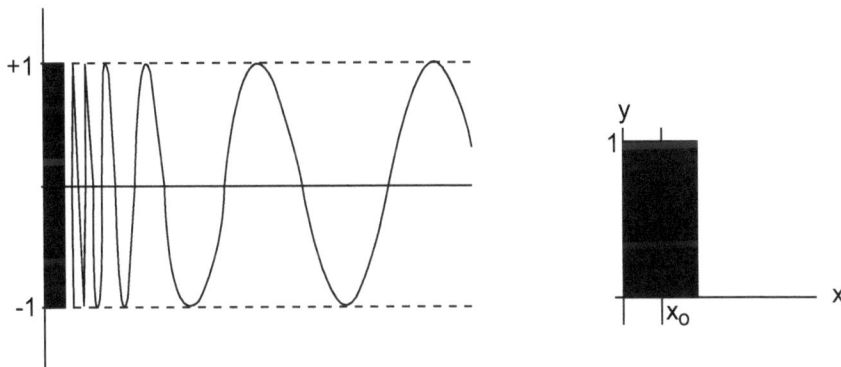

Figure 4: The topologist sine curve (left) and a rectangular function (right). The set-value of the latter at x_0 is the vertical segment $[0, 1]$.

In general, a product of two functions $f(x)$ and $g(x)$ where $-1 \leq |f(x)| \leq 1$ and $0 \leq g(x) \leq 1$ satisfies,

$$-1 \leq f(x)g(x) \leq 1; \ 0 \leq f(x)g(x) \leq 1 \text{ for } x \in \{x: f(x) \geq 0\}; \ -1 \leq f(x)g(x) \leq 0 \text{ for } x \in \{x: -1 \leq f(x) \leq 0\}. \tag{6}$$

These inequalities apply to (4) with $f)x) = \sin n\pi x$ and $g(x) = \cos^m k\pi x$ (discard β as a normalizing constant for purposes of analysis). The function $g(x) = \cos^m k\pi x$ becomes the envelope of the product rapid oscillation $(\sin n\pi x)(\cos^m k\pi x)$ and the factor $(\cos^m k\pi x)$ makes the arc at the maximum and minimum points and at the intersection with the x-axis flatter. Between the x-axis and the maximum and the minimum points the arc is steeper and closer to the vertical. This has negligible effect on the variation of the derivative and the average value of the half arc of the oscillation for large n. This observation is important for our approximations later.

When a set-valued function has probability distribution on its set-value at the point x in its domain we can compute its expectation and expectation curve over its domain shown in Fig. **5**.

We now calculate the probability distribution of the set value of the topologist sine curve at the origin by approximating it by the probability distribution of a full arc of the rapid oscillation, *i.e.*, the probability distribution of its projection on the y-axis. In view of symmetry we can use a half arc of the rapid oscillation in the interval [0, $1/4\pi k$] for the approximation.

Consider the point P on the half arc of $y(x) = (\sin n\pi x)$ over the interval [0, $1/4\pi k$] and its projection on the x-axis as the x-coordinate of P moves uniformly from 0 to $1/4\pi k$. We ask: what is the probability that its projection lies in the interval [y, y+dy)?

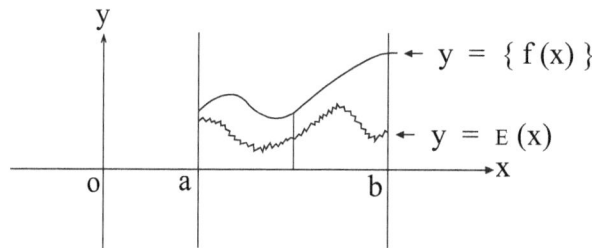

Figure 5: Set-valued function y = {f(x)} and its expectation y = E(x).

We need some physical mathematics, *i.e.*, mathematics derived from physical principle [49], such as the Heisenberg uncertainty principle, regarding the complementarity of speed and existence. We divide the interval [0, 1] into the non-overlapping subintervals [dy, y+dy) (from here on we assume that the point interval y = 1 is included in the subdivision). We ask: what is the probability that the projection of the point P in the half arc lies outside the subinterval [y, y+dy)? We apply the oscillation probability principle (a form of the Heisenberg uncertainty principle) [81] that says this probability is proportional to the speed of the projection of P on the vertical interval at the origin [−1, 1], *i.e.*, the derivative dq/dw (instead of dy/dw, where w is a dummy variable for purposes of integration). We drop the proportionality constant since it will be absorbed by the normalizing constant to the probability distribution later. We shall calculate the probability distribution in terms of derivative variation using the physical principle, complementarity of speed and existence. We compute the probability on a half arc of the rapid oscillation,

$$y(x) = \sin\pi nx, \tag{7}$$

at the origin. Differentiating (7),

$$dq/dw = \pi n\cos\pi nwdw, \quad dp/dw = (1 - \pi n\cos\pi nw)dw, \tag{8}$$

where w is the dummy variable for integration. To normalize dp/dw we note that the pre-image of the projection of the point P in the vertical interval at the origin lies in the interval [0, 1/4n]. We divide the second equation of (2) by the integral,

$$\int_{[0, 1/4n]} (1 - \pi n\cos\pi nw)dw = (w - \sin\pi nw)|_{[0, 1/4n]} = -1/4n + \sin\pi/4 \approx (\sqrt 2)/2, \tag{9}$$

since n is large. The normalized probability distribution is given by,

$$dp/dw = (\sqrt 2)(1 - \pi n\sin\pi nw)dw. \tag{10}$$

Energy conservation requires that the distribution of energy be uniform among the arcs or cycles in the case of the primum regardless of arc or cycle length (follows from the principle of uniformity of the energy conservation equivalence law). Therefore, the energy density along the full length of the photon is also uniform. Let σ be the energy density in appropriate units along the photon's axis. We find the generalized integral in the vertical interval [0, 1] at the origin and the ordinary integral along the full length of the photon to find its total energy. Moreover, since the magnitude m only affects the roundness of the envelope at its crest and not significantly the area under it, we replace m by 2. Note, further, that our computation covers one fourth of the energy of the photon. Therefore, the total energy is four times the energy under the half arc at the origin and is given by:

$$(4\sigma\sqrt{2})\int_{[0,\,1/4\pi k]}\int_{[0,\,1/4n]}\sin\pi nw(1-\pi n\sin\pi nw)\cos^2\pi kx\,dw\,dx$$

$$=(4\beta\sigma\sqrt{2})\int_{[0,\,1/4\pi k]}\int_{[0,\,1/4n]}(\sin\pi nw-\pi n\sin^2\pi nw)\cos^2\pi kx\,dw\,dx$$

$$=(4\beta\sigma\sqrt{2})\int_{[0,\,1/4\pi k]}(-(1/\pi n)\cos\pi nw\,\cos^2\pi kx)dx\,|_{[0,\,1/4n]}$$

$$-(4\beta\sigma\sqrt{2})\int_{[0,\,1/4\pi k]}(w/2-(1/4)\sin2\pi nw)C)dx\,|_{[0,\,1/4n]}$$

$$=(4\beta\sigma\sqrt{2})\int_{[0,\,1/4\pi k]}((1/\pi n)\cos(\pi/4)\cos^2\pi kx)dx$$

$$-(4\beta\sigma\sqrt{2})\int_{[0,\,1/4\pi k]}(\pi/8)\cos^2\pi\gamma x)dx$$

$$+(4\beta\sigma\sqrt{2})\int_{[0,\,1/4\pi k]}(1/2)(\sqrt{2}/2)^2)\cos^2\pi kx)dx$$

$$=(4\beta\sigma\sqrt{2})\int_{[0,\,1/4\pi k]}(1/2\pi n+(\pi/8-1/4))dx$$

$$\approx(4\beta\sigma\sqrt{2})5.64(\int_{[0,\,1/4\pi k]}0.14dx=(4\beta\sigma\sqrt{2})0.79/k\ J,\qquad\qquad(11)$$

since n is large of order of magnitude 10^{34}. The integer k is also large, although much smaller than n since the length of the photon is the period of $\cos^2\pi kx$ which is π/k. Then the energy density σ can be computed from 11 and the total internal energy of the photon.

This can be checked with the known energy of specific photon with the same parameters n, k. Then we can compute the numerical energy distribution of the photon along its axis. We can similarly compute the energy of a primum by considering the uniform energy density of its flattened projection to be concentrated on its envelope, calculating the sum along the full length of its profile and taking the full rotation of the latter suitably to find the total energy of the primum. Note that the total energy of the photon equals the total energy of the primum it comes from where the cycles convert to the arcs of the photon as rapid oscillation plus the linear momentum imparted by its carrier basic cosmic wave.

Primal Interaction

Primal interaction is governed by flux-low-pressure complementarity and the next natural law.

Flux Compatibility

Two prima of opposite toroidal flux spins attract at their equators but repel at their poles; otherwise, they repel at their equators but attract at their poles. Two prima of same toroidal flux spin connect equatorially only through a primum of opposite toroidal flux spin between them called connector.

The proton consists of two positive quarks joined by a negative quark equatorially, by flux compatibility (Fig. **3**). By energy conservation, their axis are coplanar; its charge: 2/3 – 1/3 + 1/3 = +1. This means that there is net coherent vortex flux around the proton.

By flux compatibility the electron can attach itself to a positive quark of the proton at any point but energy conservation and optimality of the energy conservation equivalence require that it attaches to both +quark beside the negative quark as the most stable position but pushes the negative quark a bit by flux compatibility so that their centers viewed from the north pole form the vertices of a quadrilateral. In its interior are the coherent vortex fluxes of the positive quarks, negative quark and electron that make it a region of low pressure or depression. By flux-low-pressure complementarity its interior sucks neutral primum around it since charged primum is repelled by primum of the same charge already in the coupling. Therefore, only suitably light neutral primum fits in and that is the neutrino. Thus, we have just composed the neutron consisting of a proton, electron and neutrino. Its charge is: $+2/3 - 1/3 + 2/3 - 1 + 0 = 0$, *i.e.*, neutral, and there is no net coherent vortex flux around it. The vortex flux of a coupled primum is also discular for the same reason as the simple primum's is due to greater centrifugal force at the equator.

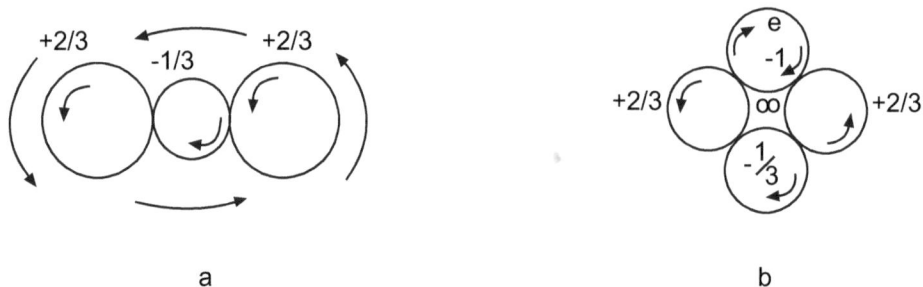

a   b

Figure 6: The proton: two +quarks joined by a –quark at their vortex fluxes' rims by flux compatibility (left). The neutron consisting of a proton, electron and neutrino (right); the neutrino is represented by a figure 8 since it is a coupled primum of opposite but numerically equal charge joined likewise at their fluxes' rims.

Since the masses of the neutron, proton and electron are known [52] we compute the mass of the neutrino.

Neutron: 1.674×10^{-27} kg

Proton: 1.672×10^{-27} kg

Electron: 9.611×10^{-31} kg.   **(12)**

Converting to atomic mass unit (amu) we obtain their masses:

Neutron: 1.0087 amu

Proton: 1.0073 amu

Electron: 5.486×10^{-8} amu   **(13)**

and the mass of the neutrino:

$$\eta = 8.5 \times 10^{-8} \text{ amu or } 1.55 \text{ times electron mass.} \qquad \textbf{(14)}$$

It was thought for a long time that the neutrino had no mass which is impossible since it is matter but it is still a subject of hot pursuit [99].

Such calculations and qualitative modeling belong to quantum algebra (see [62 – 68] for quantitative models of the superstring).

The Atom

The nucleus consisting of protons alone is the first to form in the atom. Their toroidal fluxes add up and provide a coherent vortex flux around it which is discular, thick at the nucleus and thin at the rim like the gravitational flux of a galaxy, its dual (traced by its visible halo of stars, dust cloud and other visible matter). The eye is a region of calm but since the atom, like a primum, has spin (rotation as a unit physical system) some forces come into play like centrifugal force that affects the arrangement of the nucleons. When there is only one proton in the nucleus it coincides with the eye of its vortex flux. If there are two they are joined equatorially by a negative quark in the eye of their combined vortex flux; if three they form an equilateral triangle, by centrifugal force and energy conservation, and are joined pair-wise equatorially by negative quarks. For nucleus with more protons they form rings, their common axis normal to the equatorial plane at the center each ring joined pair-wise equatorially and polarly between rings as much as possible. By energy conservation, there is a threshold of number of protons in a ring so that beyond it another ring forms, since a single large ring is unstable and violates energy conservation and energy conservation equivalence. The threshold for the number of protons in a ring can be determined experimentally. Large atom like uranium has several layers of such rings so that their profile viewed from the equatorial plane is similar to that of a primum, sinusoidal of even power. This arrangement of the nucleons conforms to the combination of energy conservation, energy conservation equivalence (e.g., optimal symmetry, universality of oscillation and related arrangements) and centrifugal force imparted by the spin of the nucleus where the rings lie at the inner boundary of the eye. They form several strings of prima joined polarly depending on the geometry of the equally spaced strings of protons joined N to S or S to N around the axis stretching from pole to pole and forming a bulge of sinusoidal profile of even power like that of a primum.

The charges of the protons add up to the charge of the atom so that their combined charge is equal to the number of protons in the nucleus. However, the nucleus itself is neutral since the protons' induced toroidal fluxes cancel each other's charges there. As positive coupled primum, the nucleus is a magnet of positive polarity with the vortex flux around it providing the magnetic field. With the right-hand rule and index finger pointing in the direction of the vortex flux the thumb points to the north-pole. Viewed from its north-pole the vortex flux of a free atom spins counterclockwise so that by Newton's action-reaction law the nucleus rotates in the opposite direction.

By flux compatibility and flux-low-pressure complementarity, electrons are attracted to the nucleus away from the eye but being light it is swept into orbit by its flux. A stable atom has orbital electrons equal in number to the protons in the nucleus. Otherwise, it is an ion, positive (when there is a deficiency in orbital electrons) or negative (when there are more orbital electrons than protons. The most stable elements are inert. Moreover, by Flux-low-pressure complentarity non-agitated superstrings steadily accumulate in the center of the nucleus; they are the principal source of energy in nuclear fission.

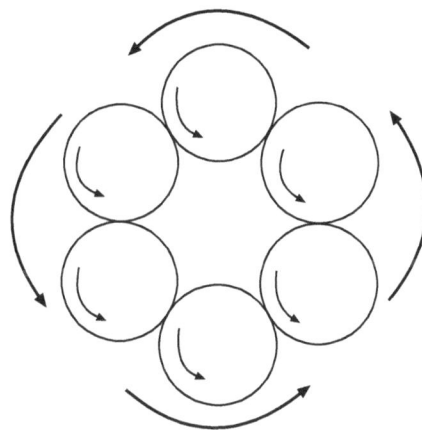

Figure 7: Light nucleus viewed from the north-pole. The –quarks that join the protons are not shown as the –quark is 1/400 as massive as the proton. In reality the protons are joined by the –quarks at the rims of their vortex fluxes, by energy conservation. The neutrons are not indicated as they do not contribute to the vortex flux of the atom; moreover, they are neutral (no net vortex fluxes around them).

Formation of Heavy Isotope

By flux-low-pressure complementarity, the neutron is the only primum sucked by the atomic eye to form heavy isotope. The chance that a neutrino is sucked by the eye is quite remote since it whizzes by at the speed of light being carried by basic cosmic wave. The number of neutrons sucked depends on the holding energy of the atomic vortex flux provided by the protons. The periodic table shows that the nucleus can hold slightly more neutrons than protons except in the case of the lightest isotope of hydrogen that has only one proton in the nucleus. Hydrogen has three known isotopes with 0, 1, 2 neutrons and possibly a third neutron that has not been discovered yet. The most common in the atmosphere has one proton alone. The one with one neutron is deuterium. Two atoms of deuterium combine with one atom of oxygen to form a molecule of heavy water. The one with two neutrons in the nucleus is called tritium.

Spin was introduced to distinguish the quantum numbers between two electrons in a sub shell which are given spins of 1/2 and −1/2. However, physicists also assign spin to other charged elementary particles; this means that spin is more than just a filler to pin down quantum number. It has physical significance; it is the rotation of a charged elementary particle as a unit distinct from the toroidal flux spin. When the primum rotates as a unit in the direction of its toroidal flux the spin is designated as positive, negative otherwise.

How are the neutrons arranged in a heavy nucleus? They pile up evenly on both sides of the equatorial plane. By centrifugal force, energy conservation and energy conservation equivalence, the neutrons of heavy isotope of an atom pile up along and on opposite sides of the nuclear equatorial plane arranged symmetrically with respect to it as much as possible. Then the complete profile of the nucleus consists of a string of protons along a sinusoidal envelope of even power with the neutrons just beyond the protons away from the axis on both sides of the equatorial plane. By centrifugal force the nuclear center is hollow along the sinusoidal axis of the proton rings. We call this the inner eye (within the proton rings).

In the immediate vicinity of the eye superstrings are also sucked, by flux-low-pressure complementarity, and accumulate in the inner eye since, being dark, hence, weightless, they are unaffected by centrifugal force. They form a mini black hole. In fissionable nuclei they convert to prima and photons released in nuclear explosion. Thus, the energy of nuclear explosion comes not from the splitting of the atom *per se* but from the agitation and conversion of superstrings in the nucleus to visible matter and energy – prima and radiation.

The electrons being negatively charged are attracted to the vortex flux away from the eye, by flux compatibility and flux-low-pressure complementarity. However, being light, they are swept into orbit by the vortex flux and become the orbital electrons of the atom. By centrifugal force, the most energetic orbital electrons are those closest to the equatorial plane. They form the outermost orbital shells. The least energetic ones cluster near the poles and form the lowest orbital shell. A sub shell may consist of one or two electrons; if there is one it has spin +1/2 and if two the second has spin −1/2 so that those belonging to the same sub shell have opposite spins. Their orbits, however, are on opposite boundaries of the vortex flux due to centrifugal force and symmetry. This completes the qualitative model of the atom.

When orbital electrons are expelled by turbulence or cosmic waves the atom becomes a positive ion. It can also become negative ion when more electrons are drawn into orbit by flux-low-pressure complementarity in addition to flux compatibility.

In a ferromagnetic material, e.g., iron, the atoms can be aligned and joined north-to-south or south-to-north poles to form a string. Then a string can be joined equatorially with other strings to form a bundle, the maximum number of strings depending on the material, by quantization. In ordinary magnet each bundle determines a line of force whose induced vortex fluxes are subject to the right hand rule. Both the north and south poles of a bar magnet have little poles corresponding to the lines of force. This can be verified by ocular inspection. The coherent vortex flux of the bundles provides its magnetic field. Iron is ferromagnetic. To make, say, a bar magnet, one wraps a coil around it connected to a direct current. Then another coil is wrapped over the first connected to alternating current. Then the second coil shakes the atoms, the first aligns them polarly to form strings and bundles and the iron bar becomes a magnet. A compass needle is made by simply rubbing it on and parallel to a bar magnet of opposite polarity. The needle becomes a magnet of opposite polarity. (The description of the magnet provided by this paragraph is familiar to the general reader and does not need any illustrations)

Prima that form atoms are converted in great numbers from superstrings in the Cosmos and in the cellular membranes of living organisms that form their cells, tissues and chemicals through the genes' brain waves [31, 76, 84]. They also form carbon that replenishes the supply in oil fields and they are continually produced in the atmosphere as well (this has bearing on the issue of carbon dioxide concentration there). In the Cosmos superstrings are converted to prima that get entangled in cosmological vortices and collect as cosmological bodies. The micro components of turbulence (to be discussed later) at the inner core of a cosmological vortex similarly converts superstrings to prima [39].

In molecular formation the valence electron serves as connector between atoms at their outer sub shells, by flux compatibility. Some atoms have weak ionization energy that the outer or valence electrons are easily knocked off by cosmic waves and become free. This is true of malleable materials like metal that is endowed with free electrons making it good electrical conductor.

Among gases the most energy conserving arrangement is diatomic, *i.e.*, coupled pair of atoms of the same gas. Ions are unstable because they interact with other atoms. It follows that neutral clusters including neutral atoms is the most stable clustering and this is the basis of chemical replenishment including that of oil.

Clarification of Quantum Phenomena

We clarify issues and puzzling and misunderstood natural phenomena of physics. The topics are not necessarily directly related to each other but they are important and need clarification.

Primal Polarity

The core of the Earth is the collected mass around its eye including the atmosphere. Technically, the Earth includes its gravitational flux. The Earth's gravitational flux shields Earth from massive asteroids and separates positive prima from their negative anti-matter that otherwise would create instability. It also shields Earth from charged particles rushing towards it from outer space.

The Earth's gravitational flux goes from West to East, is fastest at the Equator and slows down to 0 at the extremities of the Earth's eye at either Pole. Naturally, its pull on the Earth's material also slows down from the Equator to either Pole. This polar lag and its pull on a cosmological body's material was discovered in the Sun where there is a similar lag; halfway from the Equator to either Pole of the Sun the lag is 30%, a constant of nature in macro gravity, by the principle of synchronization and energy conservation.

Now let us look at the primum as it pops out of dark matter. We already noted earlier that every simple primum has charge. Therefore, a primum that is neutral is necessarily coupled. Since the dark superstring has infinitesimal induced flux it has no polarity and does not interact; when a superstring is agitated and pops out of dark matter as free primum and its equatorial plane is oblique to the direction of the gravitational flux it rotates counterclockwise in the Northern Hemisphere (clockwise in the Southern Hemisphere) due to the polar lag and aligns its equatorial plane in the direction of the gravitational flux making its vortex flux and eddy in the gravitational, its optimal energy-conserving alignment. By flux compatibility, a positive primum is pushed up so that there is abundance of free positive prima in the upper atmosphere (confirmed by shower of fragments of protons smashed by ultra-energetic cosmic waves that fall on Earth [96]). Free neutral prima are oriented randomly.

Free positive ions are counterclockwise eddies in the Earth's gravitational flux; they are pushed upwards, by flux compatibility. However, being heavy, they remain in the lower atmosphere. The electron as clockwise eddy in the Earth's gravitational flux is pushed downwards, by flux compatibility. Thus, there is abundance of free electrons on the ground. Other free negative prima including the negative quarks should be abundant on the ground also but we do not know where they are and no study on the subject has been conducted.

When the voltage between the positive ions in the lower atmosphere and the electrons on the ground reaches critical level they rush towards each other, collide and explode as lightning (a bolt of lightning has the energy of one megaton of TNT).

A moving charge is electric current and the electron is known to be carrier of electric current. It is not clear if it is the only carrier. At any rate, this separation between positive and negative prima contributes to the stability of cosmological vortices like Earth.

Outside the Earth's gravitational flux primal orientation is determined by the dominant gravitational flux there; between planetary gravitational fluxes it is the Sun's gravitational flux that prevails. Outside the solar system it is Milky Way's, etc.

Thee basic prima, namely, the quarks, electrons and neutrino, emerge steadily, especially, in living things. Since they are trapped in the cells of living organisms they are not polarized and separated by the Earth's gravitation flux.

They form atoms and molecules of tissues of living things. They are converted from superstrings in the cellular membranes by brain waves radiated by the genes [30, 31, 32].

Superconductivity and Bose-Einstein Condensate

This phenomenon where there is current through a conductor without resistance occurs when the latter is cooled to 110° K or less. We note that a flux of dark superstrings is moving charge, *i.e.*, electric current, and when there are no electrons riding on it there is no collision with the atoms of the conductor; therefore, there is electric current without resistance. This is called superconductivity. When does it occur? It occurs when the kinetic energy of the micro component of turbulence in the flux is sufficiently low that no superstrings are converted to prima and the temperature of the conductor reduces its residual kinetic energy below the ionization energy so that no free electrons ride on the flux. Both requirements for superconductivity are met at temperature 110° K or less. This is another constant of nature. To summarize, superconductivity is simply flux of superstrings without prima riding on it.

Is there potential application of superconductivity for generating energy? Hardly; it is prohibitively costly to reduce the temperature of a conductor to 110° K or less and there is no place on Earth at this temperature to cool the conductor sufficiently for this purpose.

Another phenomenon related to superconductivity is Bose-Einstein condensate verified by experiments in the US. When the temperature of a material is sufficiently close to absolute zero (temperature of 1° K has been achieved by slowing down the mechanical motion of a material sufficiently, e.g., soaking it in suitably viscous liquid at vibration resonance with it) new phenomena occur that, otherwise, do not occur at normal temperature:

(1) Atomic boundaries are blurred,

(2) Atoms pass through each other without deflection but sometimes rebounding as if something hard inside collided,

(3) Superfluidity (zero viscosity),

(4) Formation of large vortices in superfluidity and

(5) Atoms merge forming a single large ellipsoidal globule that absorbs other atoms in ones and twos and by the dozen and, with a startling suddenness, remains a huge motionless blimp.

(6) The boundary of an atom is the periphery of its toroidal flux; when the latter is suitably weakened by low temperature it is blurred.

(7) In atomic collision the induced toroidal fluxes are involved so that when the latter is suitably weakened atoms can pass through each other just like a star does through the crater of a galaxy where the stars appear highly dense but actually dispersed so that the chance of collision is almost zero. However, star collision occurs though rarely and when it happens it is called supernova. In the case of atoms actual collision of the toroidal fluxes is almost impossible because of the minute size of their toroidal fluxes but it can happen in which case there could be some bouncing.

(8) Viscosity is due to the collision of atoms in the medium mediated by their induced toroidal fluxes. Naturally, when the kinetic energy of the induced toroidal flux (charge) is weakened sufficiently due to reduced temperature, viscosity vanishes.

(9) Weakened coupling due to loss of primal energy allows passive clustering of atoms, by flux-low-pressure complementarity and flux compatibility.

(10) The weaker the induced atomic toroidal fluxes the larger the passive clustering becomes, by flux-low-pressure complementarily.

(11) However, by the quantization principle of the energy conservation equivalence law, there is a threshold of concentration and suction beyond which a sudden surge occurs that evens out the pressure resulting in the huge motionless blimp (concentration).

In a stable cluster there is balance between flux repulsion and flux attraction. However, atoms in gaseous condensate experience a small mutual repulsion or attraction depending on their species. Thus, atoms of sodium, rubidium 87 and hydrogen repel their own kind. Lithium 7 and rubidium 85 atoms attract. By introducing magnetic trap, cluster of atoms can be held together even if there is slight net repulsion between them. Large clusters are formed this way. In fact, rubidium 87 and sodium can be routinely condensed to millions of atoms, as much as 20 times larger than they would be in the absence of the repulsion.

Attractive atoms like lithium and rubidium 85 exhibit this surprising phenomenon. While this attractiveness allows formation of large cluster, with magnetic trap for stability, there is a limit of 1, 500 atoms to its clustering; beyond that total attraction is so great the cluster collapses, contracts, becomes too dense and triggers explosion that spills atoms out of the trap (from 1997 issues of *New Scientist*).

(Information (1) to (5) and (7) to (11) comes from *New Scien*tist)

Matter-Anti-Matter Interaction

Two simple prima are anti-matter to each other if they are mirror image of each other with respect to a normal to their common equatorial plane between their equators. Having opposite toroidal flux spins a primum and its anti-matter attract each other at their equators. When they get close, the momentum of their attraction forces their cycles to overlap and their fluxes to collide leading to explosion, by flux compatibility, that throws them as photons into opposite directions parallel to their respective axis. The positron is the anti-matter of the electron. The logic of GUT says that every simple primum has anti-matter. However primal polarity and the Earth's gravitational flux separate them from each other. In a controlled environment positive anti-matter can be produced on the ground and interact with its anti-matter to mutual destruction.

What about a coupled primum? Does it have anti-matter? When it has suitable symmetry it does. For example, since the proton is linear, *i.e.*, the two positive quarks and the negative quark that join them have coplanar axes, it can have a coupled primum as anti-matter consisting of two negative quarks joined by a positive quark since they are mirror image of each other with respect to a plane between their common equatorial plane. Then they are attractive; when they get close the momentum of their approach forces their cycles to overlap and mutually destroy each. However, the Earth's gravitational flux separates them and minimizes such occurrences. Neutral prima, of course, have no anti-matter. It is possible that simple prima and the protons in particular have anti-matter elsewhere in our universe. However, individual components of a coupled primum have their respective anti-matter that can mutually destroy each other.

Wave-Particle Duality

The wave-particle duality of quantum physics applies to both the electron and the photon. That the electron is a particle is well established. That it is a wave is supposedly confirmed by shooting a beam of electrons through slits in a thin plate where the beam becomes waves emanating from the slits on the other side of the plate [81], an electron in the beam supposedly becoming a wave.

The notion that a particle is both a wave and particle is a contradiction. A particle is an autonomous physical system independent of any medium while a wave is suitably synchronized vibration of the medium that projects the appearance of linear motion (analogous to the linear "motion" of neon lights due the synchronized switching of their switches). That medium is unknown in conventional physics; therefore, this contradiction cannot be resolved there.

With dark matter as the medium we can resolve the contradiction. Recall that the electron is a bulged segment of a semi-agitated (hence, visible) superstring formed by a sinusoidal helix (rapid spiral with sinusoidal profile; see profile of the primum in Fig. **2**), where the helical cycles are infinitely close and its toroidal flux, a non-agitated superstring, travels through the cycles uniformly at the speed of 7×10^{22} cm/sec [4]. However, when the electron is in flight and rides embedded between the full arcs of a pair of parallel basic cosmic waves (Fig. **3**), its carrier and envelope, the sinusoidal helix flattens into rapid oscillation due to dark viscosity with the arcs which are infinitesimally close together and the stretched projection of the helical cycles to retain the toroidal flux speed of 7×10^{22} cm/sec, by energy conservation, becoming a thin solid figure of sinusoidal shape that gives it particle properties. At the same time, the embedding pair of sinusoidal arcs of the embedding waves that serves as its envelope gives it wave characteristics.

The solidity of suitably close toroidal fluxes is analogous to the solidity of an ordinary object, say a piece of iron. All that we have in the latter are toroidal and vortex fluxes of superstrings. In fact, the atomic nuclei which consist of toroidal fluxes are so far apart that at the micro scale iron is practically a vacuum.

Recall also that a photon is a primum in flight that breaks off from its loop, the helical cycles converting similarly to rapid oscillation inside the carrier sinusoidal envelope, its arcs infinitesimally close. The toroidal flux travels through the arcs of the rapid oscillation at the same speed of 7×10^{22} cm/sec, its forward flux speed equal to the speed of the carrier basic cosmic wave which is the speed of light $c = 10^{10}$ cm/sec. Like the primum in flight the rapid oscillation inside the embedding carrier basic cosmic wave gives the photon particle characteristics while its embedding basic cosmic wave envelope gives it wave characteristics. The difference between a primum in flight and a photon is that the former is always stable being a loop while the photon is only stable when the forward toroidal flux speed (forward component) is equal to the carrier basic cosmic wave speed $c = 10^{10}$ cm/sec. Otherwise, the photon disintegrates, its toroidal flux remaining in dark matter. This explains why the photon has no rest mass. At the same time, the disintegration of the photon conforms to energy conservation since its toroidal flux, the repository of its latent energy, survives and remains in dark matter.

Clearly, the electron or photon rides only on basic cosmic wave that fits it, by resonance; radiation of long wave length may not have a photon rider. Its energy comes from the energy of its generator and the synchronized vibration of the medium (dark matter) that sustains it.

Thermonuclear Reaction

This phenomenon is currently misread the reason there is no breakthrough in fusion research that leaves nil possibility for the utilization of thermonuclear energy. It is believed that when hydrogen atoms are suitably compressed (e.g., by forcing them through narrow tunnel mechanically) their nuclei merge, shed their energy in the form of heat and form heavier nuclei like helium. In the hydrogen bomb compression is supposed to be effected by the explosion of the trigger atom bomb that presses the hydrogen atoms against the bomb shell and among themselves. This understanding is at odds with energy conservation. There is no natural law that supports it. Even in the formation of heavy isotope once the protons form a nucleus no proton can join in because charged prima are repelled by the positive or negative prima already in it. To merge two nuclei would mean introducing a proton into each other's nucleus. This amounts to reverse alchemy unsupported by the laws of nature. The repulsion between two protons alone in the narrow confines of the nucleus has been estimated at 27 tons from the perspective of conventional physics [70]. Of course, such merging can be done at great infusion of energy which would hardly generate significant energy, the aim of fusion research. This scenario is more likely:

The explosion of the trigger atom bomb agitates the superstrings in the hydrogen nuclei and converts them to simple prima (initially), heat, radiation (photons) and shock waves released in thermonuclear explosion. The converted simple prima form neutrino and light nucleons like those of helium. The trigger atom bomb in the hydrogen bomb cannot be duplicated and controlled in a reactor and, therefore, the chance of a breakthrough in the utilization of thermonuclear energy is quite remote.

Unstable Elementary Particles

There are now about 200 unstable elementary particles (prima) produced mainly in the laboratory e.g., the large proton collider. Aside from the basic prima, they vanish in split second. How do we explain their short-lived appearance? The only relevant natural principles that provide the answer are non-redundancy and non-extravagance that, in effect, say nature does not need them the reason they vanish in split second. They are man-made prima produced by the agitation of the superstrings in the nucleus of the atom due to the impact of the energized proton in the collider.

The only common stable simple prima are the basic prima and their anti-matter and the neutrino. The free simple positive prima like the positron and the +quark and anti –quark are separated from the negative prima by the Earth's gravitational flux and remain in the upper atmosphere [40]. However, they can be produced in the laboratory here on Earth.

Earthlight and Balls of Fire

The interface of turbulence pressing against each other, e.g., at tectonic plate boundary, and vibrating atoms and molecules of interfacing materials generate seismic waves known to soften metal and crack or pulverize concrete. It also occurs at compressed layers of volcanic lava. They convert dark to visible matter as balls of fire around them. They hover over and around geological faults and volcanoes. Lightning in the lower atmosphere also generates seismic waves that convert dark matter to earthlights called sprites, elves, blue jets and gamma rays, in the mesosphere [88].

Brittle and Malleable Material

We summarize briefly the nature of these materials. Brittle material has no free electrons; when vibrated vigorously say, by whizzing it through the atmosphere at 12, 500 mph, e.g., Columbia Space Shuttle's return flight, the smoothness of the insulation panel of the Space Shuttle becomes rough with vibrating molecules so that there is collision between them and the molecules of the atmosphere (friction) [40]. Consequently, their valence electrons are expelled; the material cracks or even pulverizes and the heat of friction burns it. This explains one of the puzzles of the Columbia disaster [40]: the burning of the insulation panels that had been going on during the previous decades of space exploration. Moreover, brittle material is electrical insulator since there are no free electrons to serve as carrier of charge (although superconductivity may occur at suitably low temperature [40]). It is also heat insulator because heat is due to vibration of material and brittle material does not vibrate, it cracks or pulverizes. Examples of brittle material are concrete and ceramics.

The interface of turbulence between the atmosphere and insulation panels of the Columbia Space Shuttle generated seismic waves that softened or melted the malleable materials like metal due to the greater rate of expulsion of valence electrons than their rate replacement. This was the effect on the metal chassis and attachments of the Columbia and explains its separation and eventual crash, the second puzzle of that disaster.

Malleable material like metal has free electrons; therefore, it is an electrical conductor. When distorted, e.g., bent, its valence electrons are expelled. When the distortion is restored free electrons replace them. This is what elasticity means. That is why steel is used as spring; it has almost perfect elasticity. Vibration is rapid distortion and "restorsion" and only malleable materials are capable of it. By resonance vibration spreads through the material and that is heat conduction. When metal is vibrated beyond a critical point and the rate of expulsion of valence electrons is greater than the rate of replacement it may soften or even melt. Valence electrons may also be expelled by the impact of seismic waves. The softening of metal attachments of building foundations and concrete reinforcement due to high intensity seismic waves during earthquake knocks out buildings and cracks or even pulverizes concrete. It is not really the usual gentle rocking action during earthquake that causes much destruction but the impact of seismic waves. There is technological possibility here: making suitable alloy resistant to the softening impact of seismic waves and composite resistant to cracking and pulverization by them. It offers the possibility for constructing earthquake-proof structures.

To summarize: (a) brittle materials are heat and electrical insulator, inelastic and crack and pulverize on impact of seismic waves and (b) malleable materials are heat and electrical conductor, elastic and soften or melt on impact by seismic waves.

Metal Fatigue

This phenomenon is not presently understood. Bridges suddenly collapse, a plane's pylon that connects the engine to the wing falls and metallic casing of a train's wheel breaks off, gets stuck dangling under the train and gets caught at rail junction causing terrible derailment accident at rail junction. The last tragedy happened in Germany five years ago. Investigators attributed the accident to metal fatigue but they did not know the cause of this phenomenon. An earlier tragedy happened to an American Airlines plane just after take-off from O'Hare International Airport in Chicago in the 70s when a pylon broke and the engine fell off causing the plane to crash and killing all 392 passengers.

Metal is not perfectly elastic; when subjected to repetitive distortion like vibration, the net loss of valence elections accumulates over time, reaches a critical point and the metal snaps. This explains the cause of the above accidents involving repetitive motion, mainly vibration. One can experiment on it using metallic paper clip by bending it back and forth; it will eventually snap and break. This infinitesimal but progressive deterioration of vibrating metal is not detectable. Therefore, test of material should be made at appropriate conditions for its use to determine when metal fatigue sets in to avoid accident.

Brownian Motion and Pressure

Brownian motion is due to the impact of cosmic waves on the atoms and molecules of gas and liquids that throws them in all direction; combined with their collisions, they are sent into erratic motion. When confined their motion creates pressure on the container, by momentum conservation. Since the atoms and molecules of gas are light and widely dispersed the effect of gravity on pressure is minimal. However, for dense materials like liquids pressure from gravity is considerably higher that a metallic pipe at suitable depth under water can be flattened.

Carbon, Oil and Biological Species

Carbon is the basic element of most living organisms. There are four kinds of carbon: diamond, the hardest known (tempered by heat and extreme pressure and sudden cooling deep underground), coal (solid, combustible and found on or near ground surface), oil (liquid, volatile, trapped in cavities underground) and ordinary carbon, the main ingredient of living organisms.

There is the mistaken belief that carbon comes from decayed organisms of the past. This is at odds with the laws of nature. During the ascendancy phase of a cosmological body, Earth in this case, the direction of development of physical systems including biological systems is from simple to complex. The Earth is on its ascendancy phase confirmed by the receding Moon, *i.e.*, its power as a vortex is still rising and catapults the Moon outward at the rate of four inches every century. In fact, the pull of gravity on the Earth's surface 65 million years ago was 67% of the present [83]. In the 1970s it was predicted that oil in the Middle East would be exhausted in 12 years and environmentalists were excited about it. There is no sign however that it is going to happen and oil in the Middle East continues to flow and fuel wars. In other words, carbon continues to replenish itself in regions of the Earth where it has established foothold.

Carbon forms in the Earth's interior, atmosphere and cellular membrane of living organisms [76, 84] that accounts for the growth of living organisms. There is a law of nature that says: given suitable boundary conditions all possible physical configurations favored by them arise stochastically but only the stable ones survive [31]. It is also known that some crystals and pearls grow once they gain some foothold. In these cases it looks like the presence of these elements constitutes a favorable boundary condition for their expansion and replenishment. It is generally believed that water is a precondition for the emergence of biological organisms. However, there are organisms that breathe and thrive on pulverized metal such as manganese discovered two miles underground in the mines of South Africa [10]; obviously, they cannot survive with even a mist of water.

Just like biological species that can arise anywhere whenever suitable conditions are attained (after all, any physical system is ultimately made up of superstrings) and their emergence favors growth and further development so does carbon. Although carbon continues to form in the atmosphere, there is no favorable condition for expansion and further development there. Why, then, should human life emerge only in Africa? Even in Africa the existence of two independent species of homo sapient in Kenya has been established recently and yet scientists are looking for the origin of life in space and some speculate that the seeds of life were brought here by comets or meteors. The *Raelians* (a cult founded in France that has taken root in South Korea) even assert that human life was engineered by aliens from outer space. To the question of whether life exists elsewhere in the Cosmos beyond the solar system or even beyond the Milky Way, the answer, from GUT's perspective is: very likely; this applies to the question of existence of intelligent beings. After all the basic constituent of matter is everywhere.

Chances are, as in carbon-based biological species, carbon and its various forms, e.g., coal, oil and diamond, form stochastically and are distributed unevenly. Once they emerge in a particular region they grow just like crystals and pearls which are now cultured commercially. Scientists may study the favorable conditions for the huge accumulation of carbon in certain regions of the Earth.

Explanation of Puzzling Phenomena

We introduce puzzling phenomena that now have explanation by GUT. Presently unexplained phenomena are called paranormal.

Magnetic Levitation

The magnetic flux is the coherent vortex flux induced by the toroidal fluxes of the prima. The magnetic flux of a magnet is the coherent vortex flux of its prima and the magnet itself is like the nucleus of the atom. By flux-low-pressure complementarity, its eye is a region of low pressure in the Earth's magnetic flux. Therefore, it pulls the magnet to float on it. When the magnet is suitably light and balanced, it levitates. If unbalanced it wobbles erratically and plunge to the surface.

Human Levitation

Since the human body is stable as a physical system it has globally coherent vortex flux of its prima around the body, a consequence of energy conservation and energy conservation equivalence. Therefore, like the magnet the human body has vortex flux of superstrings around it, its eye a region of low pressure embedded in the Earth's magnetic flux (which is the same as its gravitational flux that pulls the body upward. When the human vortex flux is particularly intense, an endowment provided by the gene, the body levitates and, with proper balancing, does not wobble. Some athletes actually have this genetic endowment although it is a disadvantage for the runner since the kicking push on the ground is diminished. It can be an advantage for a basketball or high jump athlete. Since a gene can be acquired by training [30, 31] the ability to levitate can be acquired also.

Giving the mind a moment of silence regularly can invigorate it and improve control of body functions. In fact, in some very old ancient cultures like India, there is a legacy of training of and body control by the mind to be able to levitate. Yoga is another example of mind control. What has being an old culture to do with this genetic endowment? It enhances suitable genetic encoding just like evolutionary change in a species [30, 31].

Kerian Photography

Kerian photography allows a special camera to capture the image of a living thing, e.g., a fresh leaf or human body in full color that had been in a room before the picture is taken. The image is only applicable to a living thing like a fresh leaf. Why? Brain waves are the medium of the gene for its main function of producing the tissues of living organism in the membrane of its cells by converting dark to visible matter [31, 76, 84]. Therefore, the genes of a living thing in the room produce prima in the cellular membranes captured by the camera.

Telekinesis

By simply getting near a metallic object some individuals can lift or bend it. Since brain waves are highly energetic due to its fractal structure they can distort molecular bonding of metal to deform it. Metal can also ride on a beam of brain waves to lift. Again, some individuals are genetically endowed that their brain waves are at staggeringly high level of energy that can do this feat.

Update

The following figures are taken from [4, 51]:Very long radio waves: $10^8 - 10^4$ m; long waves: $10^4 - 10$ m; short waves: $10 - 10^{-3}$ m; infra-red rays: $10^{-3} - 10^{-6}$ m; ordinary light: 10^{-6} m; ultra-violet rays: $10^{-6} - 10^{-9}$ m; röntgen rays: $10^{-9} - 10^{-11}$ m; gamma rays $< 10^{-11}$ m. (The finest laser (directed light) has wavelength of 10^{-9} m)

Furthermore, [4] gives numerical characteristics of the proton: its vortex flux linear velocity is 7×10^{22} cm/sec (compare with light velocity of 3×10^{10} cm/sec) from which we conclude that this is the same speed as the toroidal flux of a primum; mass 1.67×10^{-27} kg; energy 0.8×10^{15} joules = 0.5×10^{27} MeV; angular velocity 4×10^{36} rad/sec; relaxation time 10^{22} sec; the diameter of the nucleus of the atom is of order of magnitude 10^{-12} cm.

We interpret relaxation time as the normal lifetime of the proton when its components disintegrate and become semi-agitated superstrings. This occurs in the so-called neutron star, the transition phase of a star towards black hole.

We classify under visible light all known or detectable electromagnetic waves down to the finest at gamma-ray wavelength which is less than 10^{-11} m. Ordinary light is visible. In GUT the boundary between visible and dark wavelength is 10^{-14} m. This is a reasonable assumption. At any rate, experimental physics will have to make refinements in the future. Whatever the right figures might be the three sub-states of the superstring will remain: agitated, semi-agitated and non-agitated.

THERMODYNAMICS

Thermodynamics in the broad sense is the generation, conversion, transfer or conduction and utilization of energy and includes electromagnetism. It brings us to the threshold of a new technological epoch of GUT technology based on conversion of latent or dark energy to kinetic or visible energy. Even today there are technologies belonging to this epoch like the magnetic train and the electric power plant. Just like the steam engine that was invented before the development of the science of thermodynamics the magnetic train and electric power plant were invented before the development of GUT.

Electromagnetism

Much of large scale electrical power generation today falls under electromagnetism. One exception is generation by photo-voltaic cells, a phenomenon in quantum gravity. We recall that quantum gravity is the dynamics of primal induced vortex fluxes of superstrings. As in previous section much imagination is required here since we are dealing with dark matter and energy.

Direct Current Generation

This subsection is quite familiar to the general reader; and we shall simply provide a general summary for completeness. Therefore, Direct current generation is mainly through the battery. How does the battery work? It consists of two electrodes, a metal which has low ionization energy, hence, free electrons like copper, and a semi-metal like lead that can absorb electrons beyond the orbital electrons in the vortex fluxes of its atoms, by flux-low-pressure complementarity, both soaked in liquid electrolyte that ionizes the metallic electrode creating free electrons. By introducing direct current through the electrolyte from the metallic to the semi-metallic electrode the free electrons are pushed to the semi-metallic electrode creating a voltage between the electrodes. The battery is fully charged when the free electrons in the metallic electrode have been transferred to the semi-metallic electrode. Since the electrolyte is a poor conductor, when a conductor connects the terminals outside the electrolyte a flux of superstrings in which the electrons ride flow through the conductor sustained by the vortex fluxes of the atoms in the conductor like a fan belt all the way back to the metallic electrode replenishing the lost electrons there. When the metallic electrode has been replenished the battery is discharged. Then it can be recharged as before. Such a battery commonly used in cars is called storage battery. Another kind of storage battery is lithium used in lap top computers but it does not need liquid electrolyte.

There is theoretically no limit on the power or capacity of a storage battery. One can connect several storage batteries in series as a unit to attain desired voltage. Then several units can be connected as a single storage battery in parallel circuit to desired capacity and power. However, this kind of electrical generation is impractical for transmission over long distances since much energy is dissipated by the resistance of the conductor.

Magnet and Electromagnet

In ferromagnetic material, e.g., iron, the atoms can be joined polarly to form strings, a magnet with the coherent vortex fluxes of the connected atoms comprising its magnetic flux or field, its polarity in accordance with the right hand rule of electromagnetism. Two strings can be joined equatorially by negative quarks to form a bundle of the same polarity. By the quantization principle the number of strings in a bundle is fixed depending on the material. In a magnet the lines of force are the bundles. In a bar magnet the bundles extend beyond the north pole, curve outward away from its longitudinal center and bend over toward and into the south pole and join their respective south poles to form loops of bundles. It has its mirror image of loops of bundles with respect to the longitudinal center. Their profiles are seen when a piece of paper is placed over a bar magnet and iron filings poured over it that trace the lines of force. Taking a single bundle and applying the right hand rule on it and moving in the direction of its north pole the index finger points to the direction of the vortex flux around the bundle as the thumb and index finger slide through it, bends left away from the longitudinal center (right on the other side of this center) and bends towards the south pole, across it and along the magnet. By energy conservation, the magnetic fluxes of the bundles form a coherent magnetic flux around them. Again, applying the right hand rule with the thumb and index finger sliding through the bundles across the north pole and curving down (left on one side of the longitudinal center and right on the other side) towards the south pole and across it through the magnet, we trace half of the almost annular profile of the magnetic flux. We note that the bundles along the magnet and those outside and parallel to it have opposite polarity so that, by energy conservation, there is a cylindrical eye around the magnet between the inner and outer flux. The magnet itself lies in the eye of its main or inner flux.

A permanent magnet is made by winding a double helix around it and passing through direct current in one and alternating current in the other. The alternating current shakes the atoms of the material and the direct current aligns them to form the strings and bundles and the whole magnetic flux.

When a coil of conductor with direct current through it is wound around a piece of metal it aligns the atoms of the metal, by flux compatibility, to form strings and bundles and, therefore, magnetic lines of force that form the magnetic flux of the metal turning it into a magnet. Here lies a kind of duality between electricity and magnetism – electromagnetic duality – that has many technological applications.

Electric Power Generation

We apply the electromagnetic duality to the construction of the electric generator. First we note that when a conductor cuts across a line of force direct current through it is produced. How do we explain it? The magnetic flux of the bundles flows through and resonates with the magnetic fluxes of the atoms of the conductor like a fan belt through the circuit when open. This is the basic mechanism for generating direct current through electromagnetism. We can amplify the generation of direct current by letting several conductors cut through the entire magnetic lines of force of a magnet. However since direct current is impractical for high power transmission, we generate alternating current in a generator.

The generator consists of a C-shaped magnet with magnetic lines of force going from its north pole at its end, say, the upper end, to its south pole on the lower end. The axis of the magnet extends through the gap between the poles forming a loop. When a conductor cuts across the lines of force between the poles the direct current reverses direction as it cuts across the magnetic axis. If we repeatedly move the conductor back and forth across the lines of force, we generate an alternating current. We can amplify the alternating current by winding the conductor many times around an armature and rotating it at high speed so that the rate of cutting across the lines of force is raised considerably. The magnet can be an electromagnet its magnetic flux provided by the generator. The armature can be spun by an electric motor, a waterfall or turbine in coal fired or nuclear generation. The electric power plant is an up-scaled generator.

The electric power plant has several advantages aside from higher power capacity. The rotating armature inevitably results in slicing of fluxes which is agitational on the superstrings that get converted to electrons raising the intensity of generated power. The voltage can be raised by a transformer and electric power transmitted at great distances with minimal dissipation of energy, the voltage reduced at desired level for the users.

It is clear that present electric power generation is GUT technology that preceded the development of GUT just as the steam engine preceded the development of thermodynamics. The magnetic train is another GUT technology that preceded GUT. It is based on flux compatibility applied to an electromagnet on the train and a fixed one on the track with the same magnetic flux spin. The magnetic flux of the latter pushes that of the former and the train by flux compatibility.

Columbia's Final Flight: Crossroad for Science

We explain why the disastrous final flight of the Columbia Space Shuttle occurred to avoid similar disaster in the future.

Our main thesis is: present space science, thermodynamics and aerodynamics are inadequate to explain the break-up of the Columbia Space Shuttle; therefore, a new theory is required. In other words, the tragedy was no one's fault but a matter of inadequate knowledge, specifically, an error in qualitative mathematics that failed to provide adequate physical theory to guide the program. This article relies on facts and findings provided by NASA and information shared by experts on the space program. Computation and measurement alone are inadequate to explain the disaster. Therefore, the analysis and explanation must be sought in their complement, qualitative mathematics and modeling.

With GUT we explain the events around the ill-fated final flight and recommend a program of theoretical and applied research, testing and design of materials and re-design of the spacecraft itself to meet the harsh requirements of re-entry into the Earth's atmosphere. Specifically, the new design must create suitable aerodynamic flow under the spacecraft, especially, the wings to reduce pressure and friction and keep the generation and propagation of seismic waves below the threshold of disaster.

Facts about the Final Flight of the Shuttle.

We assemble the relevant facts and information as basis for analysis as well as recommendations towards the full resolution of the problem; in particular, we offer an explanation of why the Shuttle broke up and crashed killing all seven members of the crew.

(1) According to NASA the flight was flawless on re-entry until ground control lost contact with it in mid-air at 207, 000 ft. above Earth. This is confirmed by tapes showing the crew doing their normal chore minutes before loss of contact. That the flight was flawless to the last minute of contact only shows that the instruments were not registering the incipient forces already building up leading to the break-up. This is unexplained by standard science.

(2) The Shuttle was hit under the left wing by a 2.5-pound foam that peeled off the fuel tank's outer insulation after take-off and stuck and burned under it. However, this was ruled insignificant by ground control and scientists on the ground before the break-up and by the investigation committee later. It is true that the foam had no mechanical impact on the spacecraft the reason it was ruled out.

(3) Engineers on the ground were concerned about a potential problem: that superheated gases might penetrate the Shuttle through the wheel compartments which did not happen. NASA and the investigation committee concluded definitively that Columbia's trouble started with the breach under the left wing, the spot hit by the foam. This is confirmed by tapes showing dark smoke coming from this spot and the heat shield panels that burned and peeled off. That there was a "gaping hole" that let through superheated gases into the fuel chamber was not confirmed either since there was no explosion; the fuel chamber simply separated from the Shuttle over California and the shuttle separated gently into many pieces thereafter. In other words, there was no hole at all.

(4) The authorities at NASA initially attributed the breach to some meteor, debris or engine part from previous flights that hit the spacecraft under the left wing but they dropped the idea later. A meteor cannot possibly hit the spacecraft underneath nor can debris from rocket of a previous flight since it cannot remain in the atmosphere for a long time; they drop to the ground once slowed down by atmospheric fiction.

(5) If the simulation by NASA were accurate, about half the left panels underneath burned but much less tiles burned on the right side.

(6) An expert on the Columbia program, Paul Fishbeck of Carnegie-Mellon University, volunteered this information 9 years before: on the average, 25 heat shield panels on each space capsule sustained burns 2.5 cm across on each during the return flight. Other experts confirmed this information about space flights prior to the shuttle Program. In other words, the burning of the insulation panels had been going on since the early phase of the space program. In fact, the space capsules that preceded the space shuttles had burns on the insulation around them.

(7) The surface temperature of the Shuttle was $37° - 40°$ C above the expected temperature of about 2, 000° C at that altitude of 207, 000 ft. and speed of 20, 000 kph. This unexpected temperature rise is also unexplained by present science.

(8) Interviewed on TV, an astronaut on previous flight said that he saw red glow during re-entry that gradually turned to white glow. Other astronauts saw a similar phenomenon in previous space flights.

(9) Tapes of the Shuttle show no evidence of explosion; the break-up was gradual starting with the big piece (fuel tank) separating from the spacecraft over California and several pieces over Texas. The

balls of fire that engulfed the Shuttle over Texas were due to burning combustible materials on account of greater air friction once the parts separated (they were like burning meteors streaking through the atmosphere). The gradual break-up with no sign of explosion shows softening of materials, especially, the chassis and other connecting malleable material much like what happens to metal attachments of building foundations during earthquake; this is also unexplained by present science.

That the panels under the spacecraft sustained much damage compared to panels elsewhere was expected but cannot be explicitly explained by standard aerodynamics. Although pressure is about even and proportional to the amount of displacement, there is slightly more pressure on the panels underneath because of the weight and thickness of the Shuttle. Moreover, the aerodynamic design sends greater airflow above the Shuttle that mitigates the pressure there, by flux-low-pressure complementarity, a principle that is not explicit in present aerodynamics. Therefore, friction and heat generated underneath were greater and the damage greater. There was evidence of damage there even before the break up. The left drag that triggered automatic maneuver (somersault) of the Shuttle to correct it reveals the initial impact of greater friction under the left wing.

Consider the following points.

(1)   Assuming that there was a little crack on the panel (not visible to the ground team) that let in hot gas it had no significant effect on the volatile fuel since there was no evidence of explosion. However, the separation occurred along the full length of the Shuttle that revealed a general softening of the chassis, metal frames and bolts.

(2)   The other heat shield panels burned at that relatively low temperature of 2040 degrees Celsius; heat generated at the breach could not have burned the spacecraft's tiles ahead since it flowed away from them; moreover, this temperature was unexpected and even in previous flights when the temperature was 2,000° C there was unexpected burning of the panels. This needs explanation.

Since there are unanswered questions we need to device a theory to provide the answers and explanation. This is the nature of theoretical research; a physical theory is devised:

(1)   To solve some physical problems,

(2)   Explain natural phenomena and

(3)   Having done (1) and (2), provide new explanation of old results and

(4)   Explore new technological possibilities to provide remedy.

The enriched theory of turbulence we build here while a subtheory of GUT is also its advancement because new natural laws are discovered. We have introduced them previously but they were discovered in trying to resolve these puzzles:

(a)   Burning of the heat shield panels at the relatively low temperature of 2040° C,

(b)   Softening of malleable materials like metal and

(c)   Significantly grater number of burned heat shield panels on the left side of the Shuttle.

There is one more question: if the burning of the heat shield panels had been happening during the previous decade of space flights why disastrous in the last flight of the Shuttle?

Theoretical Enrichment and Construction

The new questions raised by the break-up of the Shuttle require enrichment of the theory of turbulence developed previously [35] with additional natural laws that had some bearing on the crash. We consider the break-up of Columbia and its crash and related events as motion of a physical system. Then we build a qualitative model to explain why the crash occurred and, based on our explanation, make recommendations towards improvement not only of the Shuttle but also all spacecraft to avoid similar disaster. We do much more: propose research and development program to remedy the softening of malleable material and pulverization of brittle material under conditions that occurred during that flight, e.g., during earthquake.

We apply relevant natural laws of GUT the most fundamental being energy conservation and other laws that we have already mentioned, e.g., flux-low-pressure complementarity, for our analysis of the disaster. Then we identify the natural laws that revealed themselves during the disaster and use them to explain what happened. We start with two of them below, the second combines two natural laws into its components.

Existence of Threshold for Generation of Seismic Waves

There is a threshold of flux intensity, grinding and compression at interface of turbulence that gives rise to the micro component of turbulence that generates seismic waves.

The impact of the micro component of turbulence on the interfacing materials rises with flux intensity, grinding and compression. Flux intensity is proportional to the relative speed between the panels and atmosphere which was 20, 000 kph. The grinding intensity is proportional to friction, the latter to compression. Great flux intensity and friction induces the atoms and molecules to vibrate at critical frequency. The molecular vibration at the interface of turbulence is called the micro component of turbulence in this particular physical system (there are other sources of seismic waves as we shall see later), a generalized curve [35, 112, 113]. The micro component of turbulence at the interface of the heat shield panels and the atmosphere compose seismic waves [35] that propagate in all directions. Such seismic waves form a pair of reverse nested fractal sequences of waves that is most energetic (high frequency) at the interface whose macro envelope resonates with the atoms and molecules and cause cracking and pulverizing of brittle materials of the shuttle. It also resonates with the atoms of malleable materials that softens or melts them.

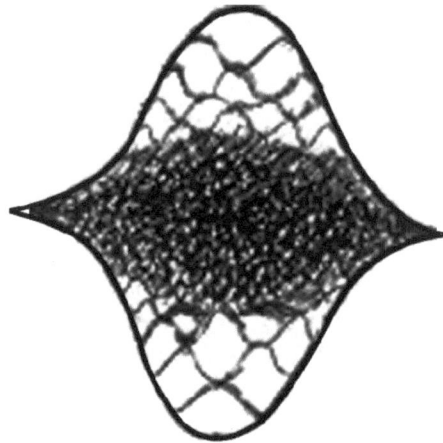

Figure 8: Two-layered seismic waves with dark component along axis generated by vibrating atomic nuclei at the interface of turbulence. Note the fractal structure of the waves from the macro envelope to the dark component (with dark frequency) at the axis or interface. The macro waves away from the axis resonate with and vibrate the atoms vigorously that cracks or pulverizes brittle materials and softens malleable materials.

During earthquake one can see the wave motion of the ground surface, the visible envelope of the micro component of turbulence, *i.e.*, the first term of the nested fractal sequences of waves whose tail terms are basic cosmic waves at the interface of turbulence. The micro component at the interface generates the fractal sequences of waves and, together, the two components generate the two-layered seismic wave one on each interfacing physical system.

In a tornado at the interface of the funnel with the ground this macro envelope picks up objects like logs; the funnel spins and catapults them (torpedo effect) into the forward direction where they hit and weaken structures before they get crushed, flattened and sucked by the funnel (this action is graphically illustrated by the movie *Twister*). Of course, this wave has no dark component and the finest wave has arc length at the atomic scale. In supersonic airplanes when a certain threshold is reached (at about 3, 000 kph) there is supersonic boom due to the macro envelope of the micro component of turbulence in the form of vigorous molecular vibration on the surface of the nose cone (forward tip) and extending through the body of the plane. This energetic vibration leads to unexplained occasional explosion and crash of supersonic jets. We re-state the second natural law obtained from the analysis of this disastrous flight. In the latter two cases the micro component of turbulence is nowhere near the critical frequency or energy that generates seismic waves.

Vigorous vibration of visible matter and hence its dark component such as in lightning, nuclear and thermonuclear explosion and compression, tension and grinding at conservative tectonic plate boundary generates seismic waves that partially convert the latent energy of superstrings to visible matter and kinetic energy like prima and photons. For example, lightning in the lower atmosphere generates seismic waves that produce earthlights high up in the mesosphere 50 to 90 km above Earth [88]. Compression and tension at geological fault produce seismic waves; so does rubbing of hot slabs of compressed volcanic lava that convert superstrings to balls of fire around volcanoes. Before and during earthquake volatile gas is released causing flame to shoot off the ground due to dark-to-visible matter conversion by seismic waves. Galactic core spin also generates seismic waves and, depending on intensity, propagates energetic cosmic waves of as much as 10^{21} ev [94]. These examples show some of the visible impact of the dark component of seismic waves.

Analysis of the Final Flight

Now there is another impact verified during earthquake: the cracking and pulverizing effect of seismic waves on brittle material and softening and melting effect on malleable materials, again, depending on intensity. The heat shield panels of the Shuttle are brittle because they are insulators. Therefore, they crack or pulverize, when hit by seismic waves, depending on intensity, and any dry pulverized material burns, even metals (e.g., iron filings poured on ember). This explains the burned panels of the Shuttle.

Seismic waves softened or melted the malleable material used in the chassis, frames and bolts (for attaching components); being malleable they are easy to mold to shape and that is the reason the Shuttle had lots of them. This explains the separation of the Shuttle that ended in a crash.

Why did the instruments not register early deterioration of the Shuttle? Most of the forces involved were dark. The indication of increasing friction on the left side was the left drag that caused automatic somersault by the Shuttle to balance itself.

How do we account for the slight rise in the temperature of the heat shield panels? The increased kinetic energy of the panel molecules was due to more intense agitation by the visible envelope of seismic waves.

Why greater damage on the left panels? There is no known force or factor that can account for the difference in effect between the left and right panels, not even the heat from the breach under the left wing since it flowed away from the panels, except the Earth's gravitational flux. The fact that there were more panels burned on the left side than on the right means that there was more resonance with and reinforcement by the Earth's gravitational flux. We have no data on the speed of the Earth's gravitational flux (in fact, we did not know it existed until the development of GUT) but as the dual of the primal induced vortex flux of quantum gravity which is 7×10^{22} cm/sec it must be considerable and in no way at the same order of magnitude as that of the panel's speed which is 2×10^7 cm/sec. However, at this order of magnitude of speed of the Earth's gravitational flux its infinitesimal effect on the panel may rise to significance, the only possible explanation for the uneven burning between the left and right panels of the Shuttle. We summarize the discussion as a natural law.

<u>*The Law of Resonance*</u>

Maximum resonance between waves, oscillation or vibration occurs when they have exactly the same characteristics including wave- or cycle-length as well as configuration. The degree of resonance declines with the difference between wave characteristics and orders of magnitude of their wavelengths. However, at suitably high order of magnitude of wavelength the infinitesimal effect of resonance by orders of magnitude nearby rises to significance.

This resonance law consisting of two components accounts for more pulverization and burning on the left side. The difference in level of resonance between the left and right panels is due to the fact that the linear speed of the Earth's gravitational flux decreases from the Equator to the North Pole where the end of the gravitational flux vortex eye is and where there is calm and no gravitational flux (the Shuttle was traveling north of the Equator going East).

The speed of the Shuttle is reckoned relative to the Earth's surface. The pull by the Earth's gravitational flux rotates the Earth and imparts the linear speed on the surface of 1, 600 kph. This is the net effect of the lag behind the pull on the Earth's crust by the Earth's gravitational flux after deducting the lag behind dark viscosity which acts on the dark component of the Earth's crust. The lag on the lower atmosphere relative to the Earth's rotation is 30 kph which is the speed of the trade winds at the Equator. At 63 km above Earth where the atmosphere is much thinner the lag must be greater so that the effective speed of the Shuttle relative to the atmosphere should be much greater than 20, 000 mph. The infinitesimal effect that rises to significance is responsible also for burning the panels on the left side albeit at lesser intensity.

With respect to previous flights how do we interpret the shift in glow around the space modules on the return flight from reddish glow to white?

Red light is long-wavelength less energetic light. White light is a mix of all lights in the spectrum from the least energetic (red) through the most energetic near the boundary of dark matter. The shift reflects the increasing intensity of the micro component of turbulence and, hence, the intensity of generation and propagation of seismic waves due to increasing pressure on the panels as the Shuttle approaches ground level.

Why did the pulverizing and softening impact of seismic waves not lead to disaster in previous flights? Let us take one step back and comment in hindsight.

The key factor in the disaster was the harsh conditions of near-space, the region within 100 km. of the Earth's atmosphere. Ignoring the breach for the moment, just consider that a little meteor of negligible displacement of air burns as it streaks through the atmosphere. Even if the pressure on the surface of the Shuttle is the same as that on the meteor, the total force rubbing on it, which is proportional to the displacement area, is at least several thousand times greater. This introduces qualitatively new complexity that is not anticipated in conventional science. In general, extreme conditions such as great speed, acceleration, compression, tension, friction and vibration introduce new complexity and bring out new phenomena unheard of under normal conditions. In the absence of adequate science there is always the chance that an unknown force would rear its head along the flight. A number of forces revealed themselves to the experts in previous space flights even in much smaller vehicles like space modules. What happened to the Shuttle could have happened in the past but we are dealing here with probability of an event that may or may not occur. Extreme conditions along Columbia's trajectory raised that probability. The generation and propagation of seismic waves at the surface of the heat shield panels set the basis for the disaster.

As harshness of conditions rises during descent including the impact of that breach and as more flights are made so does the chance of impending disaster.

At speed of descent of 20, 000 kph, even the smooth surface of the panels becomes rough and discrete with atoms and molecules and the micro component of turbulence at the interface between panels and atmosphere, ultimately attains wavelength comparable in order of magnitude to molecular size. Then the interface becomes blurred and there is resonance and reinforcement between the micro component of air turbulence and the panel molecules. This induces vigorous panel atomic vibration and collision knocking off the panel's molecular bonding (valence electrons). Since the panel has no free electrons (being brittle) there is nothing to replace the knocked off electrons and the panels tend to break, pulverize and burn. In turn, the panels' dark component resonates with and composes the basic cosmic waves into seismic waves. This is the general situation around the Shuttle whose visible impact was observed on previous flights. Another question is: why did these dynamics unravel during the return flight? The flight to the space station is mitigated by the decreasing pressure and friction as the Shuttle traversed the thinning atmosphere from the ground to the space station. They were both reversed on the return flight making the conditions doubly harsh. The effect is analogous to that of increasing speed (acceleration) that introduces additional force on the object being accelerated.

What tilted the balance of probability towards disaster was the sticky coating on the heat shield panel under the left wing from the burning foam that stuck to it which intensified the generation and propagation of seismic waves beyond the threshold for softening the malleable material of the Shuttle. The burning of the panel alone could not have caused the disaster.

Finally, we remark that the missing of the target landing by the Russian Soyuz space capsule by 400 km just about a week after the Columbia disaster may reflect the same problem that the Columbia encountered on the return flight.

Recommendations

Recovery of most of the parts and debris from the Columbia only confirmed the extent of damage and sequence of events that ensued but did not answer the many questions about the tragedy. With the enriched theory of turbulence the questions are answered. With it as a guide, we can prescribe suitable remedy for the problem of the Shuttle:

(1) Have suitable design so that the air flow would pass through the belly and under the wings of the Shuttle and buffer the interface between the spacecraft and atmosphere. This will prevent the seismic generation and propagation from reaching the threshold for softening the malleable material of the Shuttle.

(2) Find composite for heat shield panels resistant to the pulverizing effect of extreme vibration and impact of seismic waves.

(3) Find the right alloy resistant to softening under the impact of seismic waves.

If items (2) and (3) can be found then this will not only remedy the flaw in the Shuttle's design but also allow construction of earthquake-proof structures. Needless to say, testing of such materials is needed to insure reliability.

Epilogue

When the Columbia Shuttle Program was resumed after two years of suspension these problems persisted: burning and peeling of insulation panels. Disaster was averted only because the problem was fixed at the space station before each return flight. After two and a half years more of the Program the problems were not solved and it was finally terminated.

MACRO GRAVITY

This section is an exposition on cosmology based on macro gravity as a subtheory GUT and focuses on the birth, evolution and destiny of our universe. It discusses the two types of universes – ordinary and special like ours. The latter arises from a big bang, an explosion of a black hole. A black hole is accumulated mass in the eye of a cosmological vortex, its destiny.

The Cosmological Landscape

The Universe of dark matter is boundless, by flux-low-pressure complementarity [25]. Since the superstring is a nested fractal sequence of superstrings it has no beginning (no initial term). Therefore, the Universe has no beginning. Moreover, since the superstring is indestructible, it has no end. Thus, the Universe is timeless and boundless and our universe is a local "bubble" in it, a super…super galaxy 10^{10} light years across [82]. Its core is a tightly packed cocoon shaped galaxy cluster 650 million light years across discovered by French astronomers in 1994 [69].

There are two kinds of universes. An ordinary or usual universe arises from the steady shrinking of superstrings, by energy conservation that forms nested fractal sequences of depression, by the law of uneven development, and evolves to nested fractal sequences of cosmological vortices, by flux-low-pressure complementarity [42]. Its common first term is our universe itself. From there, the nested fractal sequences goes through galaxy clusters, galaxies, stars, planets, moons and cosmic dust. If we append to it the nested fractal sequences of superstrings starting with the molecules we have the whole spectrum of nested fractal sequences of physical systems from our universe all the way through the superstrings back in dark matter.

Our universe is special started by a very rare colossal event, explosion of a black hole called Big Bang, the primary premise of macro gravity; that black hole was the destiny of the core of a previous universe. The Big Bang created a super…super depression that evolved to our universe. The Big Bang was caused by suitable sequence of hits on that primordial black hole by cosmic waves that triggered the explosion.

How does an ordinary universe form? First, there is dark matter. By energy conservation, they shrink steadily and by the law of uneven development they form nested fractal sequences of depression each of which evolving to nested fractal cosmological vortices. By the quantization principle there is an optimal spread of nested fractal sequences of depression that forms a usual universe. For our purposes the basic unit of a universe is the galaxy. A galaxy consists of a main vortex and main eye in its core that collects mass around it comprising the core. It has minor cosmological vortices revolving around the core along orbits at the balance between suction by the main eye, by flux-low-pressure complementarity and centrifugal force. An ordinary galaxy has stars among its minor cosmological vortices. Some galaxies have galaxies among its minor cosmological vortices. For example, the Milky Way has 11 one of which a dead one now the Sagittarius cloud of stars [82]. Another is Andromeda with 22, all young. In a young galaxy called spiral nebula the minor vortices fall towards the spinning core along spiral trajectories due to suction by the main eye (gravity) and the effect of core spin and dark viscosity.

Only a universe launched by a big bang evolves to a super…super galaxy due to infusion of great energy by the explosion and a second but more powerful explosion called cosmic burst [37, 39, 95].

All galaxies form the usual way. The Big Bang did not form galaxies other than the super…super galaxy. There is evidence that the Milky Way formed the before the Big Bang and was drawn into our universe as the latter expanded to a super…super galaxy. The Milky Way is the oldest galaxy in its neighborhood. A young galaxy, e.g., spiral nebula, is bright with prominent spiral streamlines of minor vortices falling into its core due to gravity (suction by its eye) while the Milky Way is dim having faint spirals of falling minor vortices most of which already sucked by the main eye and joined the spinning core. Another part of the evidence is our being able to see our young universe (through Hubble) when it was only 3% of its present age. Still another is the discovery of a star in the Milky Way older than the Big Bang [97].

The Birth of our Universe

The Big Bang started our universe. In traditional cosmology it presumably occurred spontaneously with neither rhyme nor reason as if our universe emerged from spectacular violation of energy conservation. In fact, it was a natural phenomenon subject to the laws of nature.

The Big Bang created two physical systems: a super…super depression in dark matter and an expanding spherical wave front at accelerated rate called Cosmic Sphere pushed by the explosion. During this period, $0 < t < 1.5$ years, the Cosmic Sphere was compressed layer of dark matter trapped and pressed between the force of explosion and suction by the super…super depression, by flux-low-pressure complementarity, and pounded and agitated by less energetic shock waves (concentrated cosmic wave with enhanced latent energy) bouncing between its inner and outer boundaries. The agitation endowed the Cosmic Sphere and the superstrings in it with enormous latent energy. Compression kept them from conversion to prima, only semi-agitated superstrings. The more energetic shock waves pierced the Cosmic Sphere and converted dark to visible matter in the immediate exterior of the once Cosmic Sphere. The expanding Cosmic Sphere weakened and, combined with outward pressure from the compressed semi-agitated superstrings, burst at $t = 1.5$ billion years from the initial Big Bang called Cosmic Burst or second big bang [95], much more powerful than the Big Bang because of the infusion of huge latent energy by the semi-agitation of the trapped superstrings.

The Cosmic Burst released the semi-agitated superstrings that converted to simple prima at very high temperature, the first visible matter of our young universe that formed the bright radioactive clusters called quasars that peaked at $t = 2.5$ billion years [104]. Dark viscosity reduced their kinetic energy and allowed formation of coupled prima such as proton, neutron and neutrino that got entangled into usual cosmological vortices in the vicinity of the once Cosmic Sphere. This marks the birth of the early galaxies of our universe. To use a biological analogy, the Big Bang was only the mitosis of the fertilized egg and the Cosmic Burst gave birth to our universe.

The Cosmic Burst added to the breadth and depth of this super…super depression that sucked the cosmological vortices around it and formed the transitory phase of chaos [37, 39, 40] and, by energy conservation, evolved into a super…super cosmological vortex that started the evolution of our universe into a super…super galaxy. It pulled cosmological vortices along the way by gravity. The super…super depression gave rise initially to a local vortex but as visible matter formed by the agitation of the spinning collected mass, conversion of dark to visible matter by the micro component of turbulence of the spinning and falling visible matter around it that plunged into the core it imparted momentum on and raised the power of its spin. However, greater momentum and spin were added mainly and instantly by the conversion of dark to visible matter due to the micro component of turbulence at the inner-outer core of our universe. As our universe increased its spin it imparted greater centrifugal force on the galaxies but suction by the eye balanced it and induced them to form elliptical orbits around it. As its power rose further, centrifugal force surpassed gravitational suction and catapulted the galaxies outward. This explains its present accelerated radial expansion [98].

The galaxy clusters traversing our universe [103] reveals the existence of a more powerful universe that catapulted them. Such a universe could not have been formed the usual way.

There is another dynamics: as visible matter falls into the core of a galaxy *via* resonance with its dark component and pulls in dark matter, it thins out the gravitational flux and reduces dark viscosity and suction by the eye on the minor vortices. Suction reaches its peak, declines and isolates the core once again as it treks to its destiny.

Internal Motion of our Universe

One of the stunning discoveries of the last century that still haunts many physicists today is the staggering rate of radial expansion of our universe at accelerated rate [98]. Based on extensive direct measurement of the separation of galaxies from Earth, Edwin Hubble formulated his law that expresses the rate of separation of a galaxy from us at distance s from Earth:

$$ds/dt = \rho s, \tag{15}$$

where $\rho = 1.7 \times 10^{-2}$ /km distance of the receding galaxy from Earth. For convenience, we measure distance S along a great circle in the spherical dark halo of our universe. Then,

$$dS/dt = \rho S. \tag{16}$$

Since this discovery the estimate of the age of our universe increased from the original 8 billion to the present 14.7 billion and there is talk of raising it to 20 billion. Each time an older star is discovered the estimate is adjusted to accommodate it. This star-chasing game is based on the wrong premise that only our universe exists. In fact, there are others and the evidences are quite strong. One is the presence of galaxy clusters traversing our universe [103] and another is the collision of galaxies coming from different directions [105]. Galaxies in our universe travel along outward radial trajectories and cannot collide among themselves. Still another is the discovery of stars in the Milky Way older than the Big Bang [97].

Therefore, we stick to the original estimate of 8 billion to solve (4) and find the radius r as function of t. Since dS/dt = 2πdr/dt and (10) is independent of the distance between us and the other galaxy it holds when S = r. Then,

$$2\pi dr/dt = \rho r \text{ or } dr/r = (\rho/2\pi)dt. \tag{17}$$

Solving r, reckoning time from the Big Bang and taking light year and 1 billion years as units,

$$r(t) = 10^{10}e^{(\rho/2\pi)(t-8)} \text{ light years,}$$

$$r'(t) = (\rho/2\pi)10^{10}e^{\rho/2\pi (t-8)} \text{ light years/billion year,}$$

$$r''(t) = (\rho/2\pi)^2 10^{10}e^{\rho/2\pi (t-8)} \text{ light years /(billion year)}^2. \tag{18}$$

Using standard units we have, at t = 8,

$r(8) = 3.2 \times 10^{22}$ km,

$r'(8) = 840$ km/sec,

$r''(8) = 3 \times 10^{-10}$ km/secsec.   **(19)**

Since r'' > 0, our universe is on the young phase of its cycle, its power still rising. With this acceleration [98] the rate of radial expansion of our universe will eventually surpass the speed of light unless it reaches its destiny sooner. The value of ρ is based on direct observation and analysis of Doppler effect of receding light source.

Now Encarta Premium has this figure: ρ = 260, 000 km/hr/3.3 million light years, *i.e.*, the receding galaxy is moving away from Earth faster by 260, 000 km/hr for every 3.3 million light years distance away from us; it was obviously obtained from records of the past calculated or inferred from data going back at least 3.3 million years ago if they ever existed. Does it make sense?

Converting to standard units and simplifying we get $\rho = 3 \times 10^{-19}$ /km; inserting this value in (17) with the value of ρ replaced by 3×10^{-19} /km we obtain, $r'(t) = 5 \times 10^{-14}$ km/sec, the supposed rate of radial expansion of our universe, and acceleration of 3×10^{-32} km/secsec which point to a static universe that does not match present observation and measurement. Moreover, if it were correct we would have been roasted by intense heat coming from the steady formation of stars in the Cosmos, one per minute [3, 96, 102], and emergence of two baby galaxies discovered since 2004. On the contrary; the average temperature of the Cosmos remains steady at 4° C. Thus, the rise in temperature is offset by the rapid expansion of our universe [98].

Formation of a Cosmological Vortex

The core of a cosmological vortex is initially dark and isolated but the kinetic energy of its spin and the micro component of turbulence raise its temperature that agitates and converts the superstrings to prima. Thus, its expansion and enhancement of visible mass come mainly from within. As turbulence, the core's micro component generates seismic waves [35] that convert dark to visible matter in it and its vicinity; in the former mainly simple prima and in the latter also simple prima that forms atoms and cosmic dust in the cores of micro vortices and get entangled in its minor cosmological vortices. Converted visible matter in the core instantly gains momentum that augments core spin and angular momentum and, combined with momentum imparted by falling visible matter, raises the power of spin, expands its influence outward by dark viscosity and pulls and catapults outlying cosmological vortices into rotating spiral streamlines of falling minor vortices. The same dynamics is replicated in the minor vortices. Cosmic dust in a cosmological vortex also emerges from what are called cosmic ripples which are energetic cosmic waves some of which gamma-ray bursts [102]. Then it gets entangled with cosmological vortices that collect at their cores as stars, planets, moons, etc. This phenomenon of populating a cosmological vortex is dramatically illustrated by the baby galaxy discovered in 2004 and another one a couple of years later showing the spirals of visible matter just forming and falling into its core. Once minor vortices form they become self-sustaining, *i.e.*, they do what the main vortex and its core do. The spinning matter around their eyes also generates seismic waves [38] that convert visible matter within and around them.

The Earth's gravity was 67% of its present gravity 65 million years ago [83] which is roughly the same percentage of mass then relative to its present mass. The increase in mass comes from agitation by the hot spinning core by the micro component of turbulence, the principal factor in the formation of visible matter within and around it so that the Earth becomes more massive over time. This is augmented by falling matter into it, e.g., meteors and light asteroids that explode in the atmosphere. Visible matter formation in and around the Earth's core exerts outward pressure on its mantle and forces magma to ooze out of the surface, fuel volcanic eruption and pile up mountains of lava along constructive tectonic plate boundaries under the oceans that congeal into and join the Earth's crust.

Vortex Interaction

In any cosmological vortex the lucky few minor vortices that lie at the balance between suction by the eye and centrifugal force take their orbits around the eye along rotating spirals. In the solar system they are the planets and planetoids that orbit the Sun. The Sun is a minor vortex of the Milky Way and what we see is its solid core of collected mass around the eye. In an average galaxy the minor vortices are the stars and in a planet its moons if any but they are all minor vortices of Milky Way.

Consider any cosmological vortex. Since it rotates at great speed, greatest at the equator and 0 at the poles, centrifugal force throws visible matter outward at the equator. Then it becomes a thin disc of visible matter consisting of minor vortices and their cores and clouds of cosmic dust riding on the gravitational flux and thick and concentrated around the eye. The thin rim of a cosmological vortex is confirmed by the thin rings of Saturn and the other massive planets with powerful vortices that throw debris that forms these rings. The discular shape of a cosmological vortex is also seen in pictures of galaxies. The solar system is also discular with the planetary orbits along the solar equatorial plane. Mercury is its only planet that lies on its thicker portion just off the solar equatorial plane. This explains its perihelion shift of 1.67 seconds of an arc, *i.e.*, the angle that the solar radius to its center forms with the solar equatorial plane. Although the dark halo is spherical being unaffected by gravity and centrifugal force, resonance with the dark components of the visible halo along with flux-low-pressure complementarity leads to its greater concentration in the discular visible halo.

Flux compatibility and flux-low-pressure complementarity have direct bearing on vortex interaction. However, energy conservation and energy conservation equivalence are always at work in any interaction. Other natural laws are their consequences but are highlighted also because of the insights they provide in understanding the fractal principle and uneven development and resonance laws.

Spin determines interaction between cosmological vortices mediated by their gravitational fluxes by virtue of flux compatibility: two vortices of opposite spins are attractive through the common coherent induced flux at their rims along their equatorial planes; they are repulsive otherwise. If they have the same spin and their masses have the same order of magnitude, they evolve into binary vortices each revolving around the other and mutually riding on each other's spiral flux; centrifugal force prevents them from falling into each other. If they have the same spin, regardless of their relative masses, they have mutual repulsion unless one is a giant compared to the other in which case the more massive one may gobble up the other by gravity. However, if one is large compared to the other and has opposite spin, the latter rides as minor vortex or an eddy on the gravitational flux towards and merges smoothly with the core of the former unless the centrifugal force on the smaller vortex balances the main gravitational flux pressure in which case it takes elliptical orbit around the main core. Otherwise, if centrifugal force exceeds gravitational pull on a body, it may get catapulted off the vortex's influence. The galaxy clusters traversing our universe [103] reveals the existence of some powerful universe elsewhere.

Elliptical orbit, being due to radial oscillation is the most probable orbital configuration since perfect balance that yields circular orbit is unstable, by uneven development. A minor vortex along the main spiral streamline that spins opposite that of the main vortex either forms elliptical orbit around it as an eddy or gets sucked into and is crushed by the core and becomes part of it. As an eddy a vortex has relative autonomy. Two contiguous vortices of comparable masses with the same spin do not crash into each other due to mutual repulsion of opposite fluxes. Here, again, we see quantum-macro gravity duality.

As in a game of chance, an even game is unlikely over a period of time. While a pair of vortices may have initially the same mass and vortex power, once one vortex gains advantage, by uneven development, it builds up over time until it is more massive than the other. Then one becomes a minor vortex of the other. Thus, the most likely configuration of nested fractal sequences of vortices is one with a single large core vortex and many minor vortices of diverse masses along its rotating flux spirals. There are, of course, binary stars and that form when the balance is attained at the tapering of their increase in mass.

Among the intriguing questions arising from this theory is the possibility of tampering natural object to break global flux coherence and quash its capability to exert gravitational pull on other objects. (Local flux coherence cannot be eliminated since every atom has it) Moreover, by flux-low-pressure complementarity, such tampering cannot shield objects from the gravitational pull of another. However, like the stealth bomber that breaks coherence of reflected radar beams to evade detection, a sufficiently tampered body, e.g., debris like asteroids, may lose global coherent fluxes that, while acted upon by gravity, may no longer exert gravitational pull or push on other bodies. They are bodies that have lost cosmological history. To verify, we use some natural laboratory: the asteroid belt along the orbital corridors of Jupiter, Neptune and Uranus [85, 92, 108] (there should be asteroid belts also along the orbital corridors of the other powerful planets).

The irregular shape of asteroids and the objects that form the planetary rings reveals lack of cosmological history, meaning, lack of coherent gravitational vortex flux; they are debris rather than matter collected at vortex cores. They do not form gravitational clusters either, that is, they do not exert gravitational pull or push among themselves and yet they have masses. They resolve the above question and at the same time serve as counterexamples to Newton's law of gravitation.

Remember Galileo's amazement about his discovery that the rate of acceleration of a free-falling body above Earth is constant regardless of mass? This question is enlightened by a simple experiment. In a water vortex, say, a sink full of water with objects of different weights floating on it; release the water through an orifice at the center-bottom of the sink. A vortex will form and the floats will be accelerated at the same rate along spirals towards the orifice. In Galileo's experiments the bodies were falling into the Earth's core along gravitational flux spirals. The rate of acceleration is specific to the cosmological vortex; thus, Earth and Moon have different rates.

Recent study reveals that cosmic dust particles are oblong, confirming they have cosmological history, *i.e.*, like a planet, a piece of cosmic dust is accumulated mass at the core of a micro vortex. Its axis of rotation wobbles like the summer and winter solstices. Like Earth it has crust and mantle. It is estimated that interstellar dust constitutes one thousandth of the Milky Way's mass and hundreds of times more than the mass of the galaxy's planets [3]. This means that cosmic dust is a significant factor in mass enhancement of a planet. Cosmic dust continues to form and collect into stars at the cores of stellar vortices. While our universe is the first term of its nested fractal sequences as a super...super galaxy, the last terms of its cosmological vortex sequences are cosmic dust. By appending the molecules, atoms and superstrings we have the full stretch of our fractal universe all the way from the super...super galaxy through superstrings of dark matter.

The fractal-reverse-fractal algorithm [36] locates any vortex in our fractal universe starting from any cosmological body including cosmic dust particle where one can trace a fractal sequence up into the macro scale (reverse-fractal) and end up in the super...super galaxy; or go down the sequence at the micro scale and end up at cosmic dust.

Conventional science takes the view that these dust clouds formed during the last 1.5 billion years. GUT provides physical explanation of their existence and origin. Conversion to prima that form cosmic dust occurs all the time due to superstring agitation by cosmic waves, cosmic ripples and γ-ray bursts [96, 102]. However, agitation by the high temperature at cosmological vortex core and micro component of turbulence are the principal generators of prima in and around its immediate neighborhood.

"Cannibalistic" Activity of Giant Galaxies

In our neighborhood there are two giant galaxies belonging to the Constellation Virgo. Andromeda, the brightest and farthest object that can be seen by the naked eye is 2.2 million light years from Milky Way, the other giant. This is an interesting combination because Andromeda is special, *i.e.*, its initial visible matter comes from the Cosmic Burst, but Milky Way is ordinary. Both are average giants and have similar features except that Andromeda is young.

Andromeda's visible discular halo is 200 million light years across its mass equivalent to 3, 500 billion Suns [82]. It has a double nucleus or core at the center. The discular halo of a galaxy is spherical since it is unaffected by gravity and centrifugal force. However, the visible halo within it is discular in shape due to centrifugal force, thick at the center where visible matter collects due to suction by the eye but thin at the rim along the equatorial plane due to stretching by centrifugal force. Its profile seen from its equatorial plane away from the rim is sinusoidal of large even power comprised of two full sinusoidal arcs joined and tangent to each other at the ends and round but narrow at their crests. This profile is similar to the primum's [37, 43] another feature of quantum-macro gravity duality. Two of Andromeda's 22 minor galaxies are at opposite sides of and near its visible discular halo and appear headed for gravitational gobbling [82].

Milky Way contains 400 billion stars including our Sun [82]. Its visible discular halo along its galactic equatorial plane is 100 million light years across, its core or metropolis 10 million light years thick [82]. Like Andromeda its dark halo has greater concentration in the visible discular halo due to resonance and flux-low-pressure complementarity. Sagittarius, now a cloud of stars has been cannibalized by Milky Way that has gobbled some of its stars through the "saw-tooth" action by the rim of its visible halo that slices the Sagittarius' cloud of stars and throws them into a sector between the tangent and normal to the flux rim [82].

The Trek Back Home to Dark Matter

As soon as a cosmological vortex reaches its peak of power and leaves its minor vortices free, each one treks home to its destiny back in dark matter along the same cosmological path. With the thinning of its dark and visible halo the contribution of visible matter falling into the core in augmenting its kinetic energy (includes all visible energy – heat, light, motion of mass, etc.) declines but mass, spin and angular momentum continue to rise because of dark-to-visible matter conversion that introduces instant momentum to the spinning core. However, the increase in mass absorbs and puts a break on kinetic energy and agitation and reduces the rate of dark-to-visible matter conversion and energy of spin inducing steady deterioration of the kinetic energy of the primal toroidal and vortex fluxes. This results in weakening of primal bonding leading to their separation as simple prima. By energy conservation, the prima collapse to semi-agitated superstrings over a long period of time. Both the prima and semi-agitated superstrings remain around the eye due to the latter's suction.

In a star, when significant level of prima has collapsed to semi-agitated superstrings the core becomes what has been called neutron star, a misnomer since there is no such thing. Rather, the core has lost energy, specifically, the primal charge, that it has become neutral. Further de-agitation by the eye at its boundary pushes the semi-agitated superstrings to the non-agitated phase, layer by layer, and the non-agitated superstrings join the black hole in the eye. Then the core of the once cosmological vortex has reached its grave and destiny, a black hole back in dark matter. The black hole becomes naked and there is no longer suction but absence of visible matter that was sucked by its graveyard, the eye that nurtured it. Many such "voids" in the sky have been mapped and catalogued.

It is clear that in the entire history of any cosmological vortex its black hole being dark never sucks matter around it. It is the eye of the vortex that nurtures it that does. Each superstring completes a cycle (indicated by arrows): non-agitated → semi-agitated → primum (agitated superstring) → semi-agitated → non-agitated belonging to a black hole back in dark matter. The cycle may be cut short at any point and the superstring returns to dark matter as non-agitated superstring.

In a galaxy the core transitions as huge star. This was verified in 1997 with the discovery of a giant star, observed through the Hubble, 10 million times the mass of our Sun; more massive ones as much as 200 million times the Sun's mass have been discovered since then. Each star has destiny: black hole in its eye. In a massive galaxy cluster or super…super galaxy the core evolves into galaxy clusters, each galaxy evolving to its destiny as black holes. This is the scenario of evolution of the core of our universe [37, 78].

Verification of GUT in the Solar System

Explanation of natural phenomena by physical theory is part of its theoretical verification. Below is a sweeping verification of GUT in the solar system.

(1) The shielding effect of the Earth's gravitational flux, by flux compatibility and momentum conservation, accounts for the rare hit by massive asteroids on Earth (the last big one known occurred 65 million years ago; the demise of the dinosaurs is attributed to it) despite millions of them that whiz by Earth annually the latter being close to the asteroid belt in the orbital corridor of Jupiter [92, 108].

(2) An approaching heavy asteroid (along SEP) is pulled and deflected by the Earth's gravitational flux as it enters the rim, by flux-low-pressure complementarity. If it misses the narrow injection angle of about 2 degrees [38] it will be tossed by its own momentum past Earth and miss it completely. If it approaches the Earth's gravitational flux on the other side at normal speed of at least 25, 000 mph it is deflected away, by flux compatibility.

(3) There is a threshold of gravitational flux strength beyond which the shielding effect fails. This is shown by the high frequency of asteroid hits on such powerful planets as Jupiter. In the 1990s

asteroids from the tail of a dying comet landed on it and caused powerful earthquakes detected by seismographs on Earth.

(4) There is also a threshold in the other direction. When the gravitational flux is too weak the cosmological vortex loses its shielding effect. This is verified by the much poke-marked surface of the Moon. At the same time, light objects have less momentum and friction and greater sensitivity to gravity which explains the meteor showers coming from tails of comets that hit Earth frequently.

(5) However, when a meteor is large enough, say, as large as a truck, the heat of friction may raise its molecular vibration and temperature and explode it in the atmosphere. This happened recently over North Somalia and in the early 1900 over Siberia.

(6) The nested fractal structure of the Sun and its planets and the moons confirms the universality of fractal as natural law of GUT; it is conceivable that some large moons may have their own moons.

(7) Large planets like Jupiter, Saturn, Uranus and Neptune are gaseous due to the powerful spin (hence kinetic energy) of their cores.

(8) The diverse tilts of the planets relative to SEP points to the relative autonomy and independent bearing of the minor vortices as eddies; another evidence is the opposite direction of four moons of Jupiter relative to the rest of its 63 moons; they are eddies embedded in Jupiter's gravitational flux [85]; they belong to two different eddies in it one inside the other (double-layered vortex sometimes appears in kitchen sink and toilet bowl).

(9) Pluto has the most elongated orbit among the planets that lies partly inside Neptune's and Uranus' orbits [85, 101]). It has all the qualifications of a planet: it is collected mass around the eye of some cosmological vortex as shown by its spherical shape and spin; it is a minor vortex of the Sun since it revolves around it (once in 88 Earth-years [85]) along with the huge asteroids and planetoids in its vicinity [85]. It has even a moon, Charon, half its size. Where do we place the cut-off point between planet and planetoid? Pluto's disqualification has little scientific basis. This information about Pluto's orbit is quite significant because it reveals that at some point Pluto was close to the Sun but catapulted back by the latter's gravity into elongated orbit like the comets were. Another conclusion is that Pluto was a falling minor vortex of the Sun that just missed it but did not get close enough to sustain damage. In the evolution of a cosmological vortex the first to be saved from gobbling by the eye's suction are at the periphery of its gravitational flux; they are the latest to achieve elliptical orbit around the core of the main vortex. This explains the abundance of planetoids far from the Sun. At the same time, minor vortices including planetoids close to the main core, *i.e.*, the Sun, are the first to be gobbled up. That is why there are no planetoids in our vicinity being close to the Sun.

(10) The discular shape of the visible solar halo along SEP where the planetary orbits lie and the planetary visible halo traced by the thin rings of Saturn, Uranus, Neptune and Jupiter [85] are fine debris from collisions that flew off the large planets by centrifugal force or tails of comets caught by the planets' gravity or a combination of both.

(11) The asteroid belt in Jupiter's orbital corridor and along Neptune's orbital corridor [92, 108] and the planetary rings of the massive planets are major verification that bodies without cosmological history do not have gravity; they do not form gravitational cluster.

(12) The two-layered seismic waves generated by the turbulence of the spinning mass at the core of a cosmological vortex that convert dark to visible matter in this vortex are the same seismic waves generated at the interface of turbulence on Earth, e.g., tectonic plate boundary. They are known to soften metal and crack or pulverize concrete [35, 40]. This also occurs at compressed layers of volcanic lava. They convert dark to visible matter as balls of fire around them. They hover over and around geological fault and volcanoes. Lightning (explosion) in the lower atmosphere also generates seismic waves that convert dark matter to earthlights, e.g., sprites, elves, blue jets and gamma rays, in the mesosphere [86].

(13) The tidal cycle reveals an error in both relativistic and Newtonian physics. Both theories predict that it would be high tide when the Sun and the Moon are overhead presumably due to their combined gravitational pull on the ocean. This is not borne out by observation. In fact, it is low tide at this relative position of the Sun, Earth and Moon since they have the same spin and, therefore, the Earth's equatorial gravitational flux has opposite direction to that of the combined fluxes of the Sun's and Moon's fluxes on this side of the Moon. Consequently, they are repulsive and the Sun's and Moon's

gravitational fluxes push the ocean down to a low tide. Incidentally, rural fishermen can predict the occurrence of low tide during the lunar cycle from the position of the Moon. They cannot be wrong here since their livelihood is linked to the abundance of fish trapped in ponds and springs on the ocean bed during low tide.

(14) Like water eddy the hair spin at the cowlick on one's head is determined by the gravitational flux lag from the Equator to the Poles. Since the hair sticks out of the scalp at the cowlick it is the tail end that spins around. In the North there is greater counterclockwise spin than clockwise; it is the reverse in the South. An informal survey reveals that in St. Petersburg 75% of the hair's tail end spin counterclockwise around the cowlick; it is about the same percentage of clockwise hair spin in Sydney.

(15) Informal survey also reveals 100% counterclockwise and clockwise water spins in the two cities, respectively, when water kitchen sink is being emptied and toilet bowl flushed. When there are two kitchen sinks with connected bottom orifices joined into a single outlet pipe they form vortices of opposite spins in accordance with flux compatibility. It can be expected, however, that the manner by which the outlet pipes are angled with respect to the vertical can affect the direction of vortex spin.

Celestial Spectacle

We look at some amazing cosmological phenomena.

Ultra-Energetic Cosmic Waves

Cosmic rays of energy level as much as 10^{21} eV have been reported recently [94]. Traditional theories require them to be heavy elementary particles, possibly protons, coming from outside a 100-million-light-year radius from Earth. The estimate is based on supposed absence of possible source within that radius from the perspective of traditional theories. Acceleration of material object to great speed e.g., proton, is possible through centrifugal force imparted by the powerful spin of some galaxy. If some stars are catapulted so are protons. However, charged particle encounters resistance in flight and the neutron is too heavy to ride on cosmic wave and cannot be sustained at great speed by the natural vibration of dark matter. These energetic cosmic rays are not necessarily particles but cosmic waves that pack huge latent energy through their fractal configuration. They are known to smash protons in the mesosphere [94] that, incidentally, confirms our prediction that positive prima are pushed high up by the Earth's gravitational flux [40]. Such cosmic waves could have come from the cores of powerful galaxies. Then there are energetic gamma-ray bursts coming from distant regions of the Cosmos and some scientists theorize that they are due to black hole explosion.

Jet Outflow

Many interesting dynamics are displayed by galaxies and stars. Among the spectacular dynamics in the Cosmos is jet outflow of hot gas or pure prima from the cores of nascent stars and galaxies [10]. Jet outflow was the first known case of matter speeding as fast as 75 times faster than light [10]. How do we explain it? The eye of a vortex is cylindrical and normal to the plane of its discular halo along the equatorial plane, much like that of the tornado or primum. The rapid spin of the core (pure prima) around the boundary of the eye (event horizon) builds up tremendous kinetic energy and accumulation of hot gas or prima that must find a soft spot to escape through and that is the eye itself. Thus, jet outflow pops out of the eye of a young galaxy or star in opposite directions at great speed as much as 75 times the speed of light [10]. As the accumulated mass at the core cools down, the eye extremities suck the mass inward leaving spherical mass slightly flattened at the poles. This is quite evident in the Earth's flattened polar region (its curved rim seen at Lookout Restaurant near Sydney).

Opposite Orbits of Jupiter's Moons

12 moons of Jupiter revolve opposite to the revolution of the rest of its 63 moons [85]. We interpret Jupiter to have a double vortex spinning in opposite direction one of them containing the 12 moons. Such double vortex is common in kitchen sink and swimming pool when being emptied.

The Solar System

We briefly discuss the Sun as a minor vortex of the Milky Way, an average star; its solid core that we see has diameter of 889, 000 miles. It has nine planets as minor vortices; the planetary profiles are taken mainly from [85].

Mercury

Being the nearest planet to the Sun, it lies on the thick side of the Sun's discular visible halo which accounts for its perihelion shift of 1.67 secs. Since it is a light planet its core spin is weak, its solid core comprised mainly of rocks.

Mercury's Profile

Average distance from the Sun: 36 million miles; diameter: 3.032 miles; revolution around the Sun: 87.97 Earth days; Number of Moons: 0; rotation on axis: 58.65 Earth days; tilt of axis relative to the solar equatorial plane (SEP): 0°; surface temperature: −300° F to +800° F.

Venus

Venus is about the same size as Earth but has much higher temperature on the surface: 900° F during the day. The atmosphere is very dense and the planet has thick red clouds that maintain the terrible heat and rain down sulfuric acid. Its thick carbon dioxide atmosphere helps to heat its surface. The clouds of gas let in heat energy from the Sun but keep the heat from escaping. This is called the *green house effect* because it traps heat energy the same way as glass in a green house. Venus has opposite spin to Earth's; therefore, these two planets attract each other. If Venus were only close to Earth it would have opposite effect on Earth's than the Moon's. On Earth it is low tide when the Moon is overhead because they have the same spin and, therefore, mutually repulsive gravitational fluxes. With Venus it would be high tide on Earth if it where overhead.

Venus' Profile

Average distance from the Sun: 67.2 million miles; Diameter: 7, 521 miles; revolution around the Sun: 224.7 Earth days; tilt of axis: 177.3° from SEP; rotation on axis: 243 Earth days; average surface temperature: 867° F; number of moons: 0.

Earth

Earth offers much verification of GUT: from the behavior of falling objects through the spin of vortices formed by water escaping underneath the sink or swimming pool, wind spin around the eye or typhoon or tornado and the hair pattern around the cow lick on one's head. They confirm the lag in the gravitation flux from the Equator to the poles: counterclockwise in the Northern Hemisphere and clockwise in the Southern Hemisphere. Over 70% of Earth's surface is covered by water. The Earth is the only planet in the solar system known to have life and its atmosphere protects living things from dangerous radiation of sunlight. However, the probability of the existence of life elsewhere in our universe is 1. The Earth's moon is at average distance of 238.328 miles. Its spin is the same as Earth's; therefore, their gravitational fluxes are repulsive that cause low tide when the Moon is overhead. As the Moon revolves around Earth (in 29.5 days) its rotation around the axis is such that the same side always faces the Earth directly.

Earth's Profile

Average distance from Sun: 93 million miles; 7, 926 miles; surface temperature: −128.6° F to +136° F; number of moons: 1; revolution on its axis: 1 day; tilt of axis: 23.45° from SEP; rotation around the Sun: 365.24 days.

Mars

Mars called the red planets has always fascinated humans with the possibility of the existence of life, perhaps, in less intelligent forms. Since the exploration of the planet first with electronic probes and then with robots on the surface, no evidence of life has been found. There is evidence that the planet had a more vibrant past millions of years ago. Rain fell and rivers flowed but, today, dry river beds only reveal its more hospitable youthful past. Some water may still be frozen in the soil. If there were intelligent living creatures in Mars they could share some wisdom on dealing with the environment. In fact, the condition in Mars was quite favorable for igniting the spark of life: at the north and south poles are caps of water, ice and solid carbon dioxide (dry ice); temperature may reach 80° F at the equator on a summer day; at night it is colder than anywhere on Earth. Mars has the highest extinct volcano in the solar system, Mt. Olympus. The thin Martian atmosphere is mainly carbon dioxide.

Jupiter

Jupiter is the largest planet in the solar system, has diameter 11 times that of Earth – roughly 86, 000 miles – and volume roughly 1, 300 times that of Earth's. However, as a massive planet, its core spin is powerful and has enormous kinetic energy. Consequently, it is much less denser than Earth and is mainly made up of gases. Space probes so far have sunk deeper and deeper into the gases until they are crushed by the pressure or burned by the heat. Like the Sun whose inner core specific gravity is 150 (a constant of nature) Jupiter's deep interior is mainly pure prima so packed it is solid. Large as the planet is, its rotation is much faster than Earth's, confirming the tremendous power of its core spin. It makes one rotation every 10 Earth hours. Winds at the equator travel more than 100 miles per hour. The hurricane, Great Red Spot, is nearly 25, 000 miles wide (roughly the Earth's circumference) and has raged for at least 300 Earth years.

Jupiter's Profile

Average distance from Sun: 484 million miles; diameter: about 86, 000 miles; revolution around the Sun: 11.86 Earth years; rotation of axis: 9.9 Earth hours; atmospheric temperature: −227° F; number of moons: 63; tilt of axis: 312° from SEP.

Saturn

The most distinctive feature of Saturn is the thin layer of rings around it made up of particles of ice each orbiting around it like tiny moons (Jupiter has also slight rings of this type). They serve as tracer of the planet's discular visible halo (concentrated dark halo) due to the full effect of centrifugal force exerted on them along Saturn's equatorial plane by its core spin and dark viscosity around it. Saturn's rings are 25, 000 miles wide and about 100 feet thick. Aside from these rings there are also 18 moons orbiting the planet most of them very small but Titan is the largest in the solar system. Saturn has thick atmosphere containing the kind of chemicals that might have helped form life on Earth 4 billion years ago. Saturn is a huge planet similar to Jupiter. It is surrounded by clouds of gases mainly.

Saturn's Profiles

Average distance from Sun: 887 million miles; rotation on axis: 10.7 Earth hours; revolution around the Sun: 29.42 Earth years; approximate diameter: 71, 200 miles; average atmospheric temperature: −285° F.

Uranus

Uranus looks like a green sphere without any surface feature. It is 20 times farther from the Sun than Earth and spins on its side, *i.e.*, its axis points directly to the Sun. During its 84-year orbit around the Sun one of its poles points directly at the Sun. Uranus is a giant gas like Saturn with a solid deep core of pure prima. Beneath its deep atmosphere is a giant ocean of liquid or icy water, methane and ammonia. Uranus has 17 small moons ranging in diameter from 20 miles to 1, 000 miles.

Uranus' Profile

Distance from Sun: about 1.783 billion miles; rotation on axis: 17.2 Earth hours; revolution around Sun: 83.73 Earth hours; average atmospheric temperature: −355° F; diameter: about 32, 000 miles; number of moons: 17; tilt of axis: 97.86° from SEP.

Neptune

Neptune is a hazy planet 30 times farther from the Sun than Earth. It has deep atmosphere covering an ocean of water, methane (that makes its color blue) and ammonia. One of Neptune's moons, Triton, is probably the coldest place in the solar system. It is Neptune's largest moon and has rugged terrain and active volcanoes. Nereid, another of Neptune's moons, orbits the planet at a distance of 3.5 million miles.

Neptune's Profile

Average distance from the Sun: 2, 794 billion miles; rotation on axis: 16.1 Earth hours; revolution around the Sun: 163.7 Earth years; approximate diameter: 30, 500 miles; average atmospheric temperature: −355° F; number of moons: 8; tilt of axis: 29.56 from SEP.

Pluto

Pluto is sometimes called the dark planet being very far from the Sun. It is also the smallest planet. It has one moon, Charon, half its size. They circle its other like dancing (Fig **8** paths). Being small, Pluto has weak core spin and is made up of rocks and ice. The closest point from the Sun on its orbit is 2.8 billion miles; farthest, 4.6 billion miles.

As we have already seen Pluto is in all respects a planet, the accumulated mass at the core of its cosmological vortex and a minor vortex of the Sun. Naturally, it has gravity and orbits around the Sun. Its orbit is a bit slender and elongated that part of it lies inside the orbits of Uranus and Neptune [85, 101]. It just happens to be smaller than the other planets, its diameter 1, 413 miles [85].

Beginning 1979 Pluto was closer to the Sun than Neptune but it traveled outside Neptune's orbit in 1999 and will be the most distant planet from the Sun since then for 227 Earth years. In 2006 the International Astronomy Society officially expelled Pluto from the solar system, a bit unfair since it has all the features of a planet and is richer in features than some planets.

Pluto's Profile

Average distance from the Sun: 3.675 billion miles; diameter: 1, 413 miles; revolution around the Sun: 248 Earth years; rotation on axis: 6.4 Earth days; average surface temperature: –370° F; tilt of axis: 122.24° from SEP.

Debris in the Solar System

We consider debris in the solar system which has impact on Earth. A comet is partial debris, a planetoid that just missed falling into the solar core and was catapulted back by the Sun's gravity into elongated orbit that extends far beyond Pluto but damaged by the intense gravity of the Sun as it whizzed by during its near hit on the solar core. The damaged part consisting of fragments become its tail. It is pushed away by the pressure of solar radiation so that its tail is always away from the Sun along its entire orbit. As a comet passes by a massive planet its tail is pulled by the planet's gravity and some are suspended in neutral region away from the gravitational influence of any planet but ride on the gravitational flux of the Sun and orbit around it. The comet's tail consists of asteroids and meteorites. They are the same materials but differ only in mass. Although they have mass they have broken vortex flux and, therefore, have no gravity. Clearly, they are counterexamples to Newton's law of gravity. Huge asteroids cluster in the orbital corridor around Neptune [92, 108] and smaller ones in the orbital corridor around Jupiter and beyond the orbit of Mars. They are called asteroid belts. Millions of asteroids from the latter asteroid belt whiz by Earth every year.

At the same time, planetoids concentrate at and are close to the periphery of the solar gravitational flux as they escaped being gobbled up by the core of the solar vortex, the centrifugal force on them balancing the gravitational suction by the eye. They are really planets, the collected mass around the eye of minor vortices of the Sun that range in size from the smallest planets to the tiny cosmic dust. When a comet passes by, they may be pulled out of their orbits, by flux-low-pressure complementarity, and collide with others adding to debris of asteroids and meteorites. When the comet returns the debris may be nudged from their orbits and thrown in the direction of the inner planets boosted by solar gravity.

The Asteroid Belts

The corridor between the orbits of Mars and Jupiter has been referred as the asteroid belt [85]; now we add another asteroid belt with planetoids and huge asteroids in it in the orbital corridor between Neptune and Pluto [108]. It is here where thousands of irregularly shaped asteroid orbit the Sun. Their irregular shape reveals that they are debris coming mainly from tails of orbiting comets. It contains not only asteroids but also planetoids orbiting the Sun. In the evolution of the Solar system the planetoids closer to the Sun were the first to be pulled to its inner core by gravity and those farthest from it survived. They are the planetoids in this belt.

Will Asteroid Hit Earth this Millennium?

While meteorites shower Earth by the millions daily there are only a couple of asteroid hits during the last 65 million years. Why? Consider the present configuration of the Earth and Sun where the equatorial plane of the former forms acute angle with that of the latter. If an asteroid (solar asteroid travel along the solar equatorial plane) approaches, say, on its return trip from the other side of the Sun, it is pulled by the Earth's gravitational flux, by flux-low-pressure complementarity. Viewed from the Earth's North Pole the asteroid is deflected to the left by it, by resonance with its dark component (since the Earth's gravitational flux is counterclockwise). If it traverses left of a narrow injection angle within which it would crash on Earth, it would miss Earth entirely by its own momentum. Light objects, e.g., meteors are overcome by Earth's gravity and pulled by it into streamlines of falling minor vortices and debris and crash on Earth. If the asteroid approaches right of the injection angle traveling at speed of about 25, 000 mph, it will be deflected away by the opposite gravitational flux of the Earth and miss Earth, by flux compatibility.

The chance of asteroid hitting Earth is greatest when its gravitational flux equatorial plane is normal to the Sun's. That will be at least 6, 000 years from now in the course of the Earth's magnetic flux wobble. For now that chance is quite remote. Therefore, to the question of whether an asteroid will hit Earth this millennium, the answer is: quite unlikely. Consequently, the preparation for blowing up or deflecting an approaching asteroid is premature.

The Earth

As theoretical application of GUT this section is a sequel to [38]; it provides a qualitative model of the Earth both as minor cosmological vortex of the Sun and its core, the accumulated mass around its vortex eye that extends to the atmosphere. The Earth's cosmological vortex that coincides with its gravitational flux carried its minor vortices of the past with the Moon as the only survivor that escaped gobbling by its eye. The entire vortex that coincides with the gravitational flux is called dark halo and the concentration of visible matter along the Earth's equatorial plane due to centrifugal force of spin and resonance of visible matter's dark component with the dark halo is called the visible halo. Its shape is discular, thick around the Earth and thin towards the rim that extends far beyond the Moon. This discular configuration of visible halo is uniform for all live cosmological vortices including active galaxies (the Sagittarius cloud of stars is an example of a dead galaxy [82]).

Early Cosmological History of the Earth

Like any cosmological vortex the Earth started as dark vortex, a term in the nested fractal sequences of vortices that arose from the nested fractal sequences of depression due to the steady shrinking of superstrings, by energy conservation and uneven development [42]. Then the kinetic energy of its spinning dark vortex and the impact of cosmic waves coming from all directions agitated and converted dark matter to its initial visible matter and evolved into nested fractal sequences of cosmological vortices with the initial dark vortex, now populated by its visible core (collected mass that extends to the atmosphere) and its only remaining minor cosmological vortex, the Moon, as first term or main vortex.

Just to put the Earth on the right perspective, in the nested fractal sequences of cosmological vortices that comprise the Milky Way, the Earth is a tail sequence of a minor vortex of the Sun with the Sun's vortex as first term. Therefore, it is a minor vortex of the Milky Way. As noted earlier the Milky Way was far from the Big Bang and Cosmic Burst since the latter is visible through the Hubble [96].

Global Oceanography

Since the pull by the gravitational flux is proportional to density and density varies inversely with the radius there is a layer lag in rotation between inner and outer layers of the Earth's core. For example, there is a layer lag between the Earth's crust and the ocean and between the ocean and the atmosphere.

Consider the Earth's crust and the Pacific Ocean. The layer lag between them produces an ocean current along the Equator from East to West. This current is like a river in the ocean that is as wide as 80 km at some points and splits into eddies or local current cycle that narrows it down at other points. Blocked by the Asian land mass, its northern strip curves northward towards the North Pole but blocked by the Chinese coast, it curves again eastward towards Japan and under Siberia, across the Bearing Sea and, blocked by North America, curves downward off the Mexican Coast and then curves westward to join the Equatorial current, by flux-low-pressure complementarity, and completes the Northern Pacific Ocean Cycle. The current flows in the clockwise direction.

As the cycle curves it narrows due to its momentum that presses it against the land block aside from the effect of eddies that draws away a strip of the current. Thus, there is a narrowing of the Northern Pacific Ocean Cycle from the eastern seaboard of the Philippines through east of the coast of Japan, then south of Alaska as it bends southward and, again, west of Mexico as it bends westward to complete the cycle. Just like a constricted water hose that shoots off water jet, when the current narrows its flow rate rises, a consequence of energy conservation. This happens at bends in the Pacific Ocean.

Due to friction with the seabed that pulls the ocean bottom with it towards the East, the lower strip of the west end of the westerly Equatorial current off the Philippine Coast bends downwards and curves into a horizontal current towards the East as it is being pulled by friction with the seabed that goes eastward. Then this countercurrent under the northern strip of the westerly Equatorial current goes all the way East, bends upward and feeds into the origin of the westerly Equatorial current to complete the vertical cycle (illustration in [107(b)]). Note that formation of cycles conforms to energy conservation. Current where the head has nowhere to go stops and cannot be sustained. In the case of rivers they empty into the sea; ocean current forms a cycle. The undercurrent in the Northern hemisphere shifts Northward relative to surface current segment due to the polar lag.

The Pacific Ocean cycle has a rough mirror image in the Southern Hemisphere that flows in the counterclockwise direction. In both cycles ocean eddies or local cycles serve as "ball bearings" to ease friction due to viscosity and facilitate ocean flow. In the past the ocean cycles were used by seafarers to facilitate travel and even today village fishermen use the local cycles for the same purpose.

The wind creates waves not ocean current; it can only produce waves that appear to travel but, in fact, only rides on the synchronized vibration of water molecules that stay in place. The dynamics is quite analogous to the neon lights that appear to travel due to the synchronized switching that turns on and off the lights one after another in rapid succession. It would take tremendous amount of force to overcome water viscosity and push a layer of the ocean to generate current. Wind is quite incapable of it.

The situation is similar in the Atlantic but in the Arctic Ocean the ocean current goes around the North Pole clockwise. However, in the

Indian Ocean the cycle is quite different. It reverses direction twice during the year: during winter it is counterclockwise and clockwise during summer. Why? During the approach of winter when the Equator tilts northward for the Winter Solstice the northern arc of the Indian Ocean Cycle coincides with the westward ocean layer lag. Therefore, it pushes the current westward from India to the South Africa Coast inducing counterclockwise current along the Indian Ocean Cycle. During the approach of the Summer Solstice when the Equator tilts southward, the southern arc coincides with the westward ocean layer lag. Therefore, it induces or forces clockwise current along the cycle. Again, this current reversal has nothing to do with the monsoons but the effect of the summer and winter solstices.

The Earth's Interior

We distinguish the Earth's inner-outer core from its core as a cosmological vortex and focus on the Earth's interior, *i.e.*, everything beneath the surface including the oceans and lakes since it influences atmospheric behavior, particularly, turbulence and weather change. For example, under-ocean volcanic activity forms pockets of warm water called *el niño*. When large huge contiguous *el niño*, say, as large as the Canadian landscape, heats up the lower atmosphere, it causes hurricane [35] and tropical cyclone. The Earth's interior, is separated into layers from the inner and outer cores through the mantel, crust and oceans.

The Earth's inner core is the hottest layer, 6,000° C [107(a)], and its most compact solid layer due mainly to the high temperature and kinetic energy that allows only formation of simple prima in polar and equatorial coupling [37, 43] that leaves minimal space between them. Consequently, the inner core has specific gravity of 150, the same specific gravity as the deep interior of the Sun. This information about the inner core is based on measurement of passage of seismic waves during earthquake and the fact that speed of passage is proportional to density of the medium.

The inner core wraps the eye and has common boundary with it. The outer layers were pulled by and spun by the Earth's gravitational flux during the early phase of its cosmological history through resonance with their dark component, the inner core pulled most effectively and spins with the gravitational flux at staggering speed due to its high density, effectiveness of pull being proportional to it. It continues to this day.

By energy conservation and energy conservation equivalence [42, 43] the profile of the inner core viewed from its equatorial plane away from it is a pair of sinusoidal arcs of even power tangent at the ends much like that of a primum [43]. The simple prima form strings of polarly coupled prima end to end and the strings of positive prima are joined equatorially by negative quarks which may also form polarly coupled strings depending on their distribution. The alternate coupled strings of positive and negative quarks wrap around the eye of the Earth in layers and form the inner core. The outer core wraps around the inner core in the same arrangement and profile except for the possible inclusion of coupled prima and light nucleons allowed by its lower temperature of $5,500°$ C [107(a)].

We do not know what the circular speed of the gravitational flux that pulls the core (we can assume minimal slippage because of its high density and compactness in which case we can assume it is spinning at the same speed as the gravitational flux). We have a clue from its dual in quantum gravity, where the circular speed of the atomic flux induced and powered by the protons' primal flux around the nucleus is 7×10^{22} cm/sec [4]. We do not know if the duality is faithful. All we can say is that the circular speed of the gravitational flux is huge. It is this great intensity of the Earth's gravitational flux within the inner and outer cores and effectiveness of pull that determines its magnetic pull and polarity.

The increased angular momentum and spin of the cores is enhanced principally by dark-to-visible matter conversion within the hot spinning cores that gains instant mass and momentum upon conversion and augments its angular momentum and spin although the contribution of cosmic dust that falls to the cores is quite significant, e.g., the mass of cosmic dust in the Milky Way is greater than the mass of all its planets combined [3]. The same is true of dark matter sucked by the eye (by flux-low-pressure complementarity) that joins the spinning core; it gains instant momentum upon its conversion to visible matter. The Earth's minor vortices were formed in the same way at smaller scale. At the same time, the increased spin of the Earth broadens its eye due to centrifugal force which, in turn, increases its suction, *i.e.*, gravity. It is known that 65 million years ago the Earth's gravity was 67% of the present [83]; its mass must have been proportionally small then also. Some scientists conjecture that the demise of the dinosaur was due to the inadequacy of its anatomy relative to its increasing weight brought by the increasing gravity of the Earth long before its total disappearance 65 million years ago.

The dark-to-visible matter conversion in the core and mantle expands the Earth's interior and creates outward pressure from the Earth's mantle that results in magma oozing out of the crust that builds mountain ranges along both sides of constructive tectonic plate boundaries under the oceans that become part of the crust. Some of them broke out of the ocean surface and formed islands. Constructive plate boundaries along the Pacific rim, called the Pacific Ring of Fire are responsible for much volcanic activity, some under the ocean and others find cavities and create volcanoes inland and feed them with lava. The Pacific Ocean seabed is studded with numerous under-ocean volcanoes responsible for much of weather turbulence there including tropical cyclones.

As visible matter accumulates around the inner and outer cores, mainly through dark-to-visible matter agitation and conversion by its intense vibration due to high temperature and generation of seismic waves by the micro component of turbulence there that convert superstrings to visible matter around them, the mantle shields the rest of the outer layers from the hot inner and outer cores allowing formation of more complex systems such as atoms and molecules and even biological species close to the Earth's surface. Ingredients of many biological species come from magma oozing out of the Earth's crust from the Earth's interior [107(a)]. This explains why volcanic islands like the Galapagos west of Ecuador and Hawaii have the most diverse collection of animal species [107(a)]. Here, there are cracks and holes on the plates that form hotspots where magma oozes out and creates volcanoes and feeds them with lava. It appears that most biological species form from magma as it cools to surface temperature. The Galapagos has the most diverse biological species in the world; it was here where Darwin spent the years 1831 – 1836 gathering data for his theory of evolution of biological species [107(a)].

The temperature of the Earth's interior drops rapidly in proportion to the distance from the outer core because the inner layer of the mantle although hot and in liquid form due to proximity to the hot core shields the outer layers, liquid being good heat insulator. As the thickest layer, the average temperature of the mantle drops considerably so that complex atoms form but mainly in liquid form. This is verified by the outflow of magma and melting of the crust when it subducts into it at destructive plate boundaries.

The crust is made of diverse materials including rocks and minerals cool enough to host biological species near the Earth's surface. Most species are on the surface and in the oceans but there are metal-based organisms underground, e.g., organisms found two miles underground in the mines of South Africa [10]. Relative to the rest of the Earth's interior the crust is quite thin like an eggshell. It is here where most destructive geological occurrences take place, mainly, earthquake and volcanic activity and erosions. The density of the Earth's materials decreases from the core outward due mainly due to the presence of more complex atoms near the surface; in the atmosphere large gas atoms take abundant space.

At this time, the Earth is at its ascendancy phase of its cycle as shown by the receding Moon which means that the power of its spin is still growing and the Earth itself is still developing to higher order, e.g., new biological species continue to form the existing ones well on their evolutionary advance [31].

Why should the spin of the Earth continue to rise when there are no longer minor vortices falling on it to enhance its rotational momentum (the impact of meteors and asteroid is negligible)? The main converter of dark to visible matter is the hot spinning core. As soon as visible matter is converted from dark to visible matter it acquires momentum instantly that it adds to the core's momentum and spin. Moreover, as long as the eye is still there it continues to suck dark matter that gets agitated and converted to visible matter by the hot spinning core and enhances the power of the spin. However, there is a break in that expansion. As the mass of the Earth becomes critically great it provides breaking action on its spin and everything else will begin to decline. Combined with the thinning of its dark halo the expansion of its gravitational flux will decline and come to a halt. Then the Earth will have reached the summit of its power and begins its trek down its destiny.

Movement of the Earth's Crust

The Earth's crust floats on the mainly liquid high density mantle of the Earth. It completes one rotation with the Earth's spin every 24 hours or circular speed of 1, 609 km/hr (1, 000 mph). Energy conservation requires uniform linear speed of the Earth's gravitational flux through the layers. However, there is dissipation of energy due to friction since the inner layers are spinning more rapidly than the outer layers and the layer lag due to ineffectiveness of pull by the gravitational flux on the Earth's materials of lesser density. The effectiveness of pull is proportional to the compactness of the materials and that is roughly inversely proportional to the distance from the Earth's inner and outer cores, the most compact layers. Next in compactness is the basically liquid mantle; then the crust being relatively porous is less dense compared to the mantle although it contains dense liquids like minerals which are negligible. Next to the crust are the relatively light oceans and the much lighter atmosphere.

There is a natural lag that does not entail dissipation of energy and that is the lag in displacement since each element of mass has to travel greater distance the farther it is from the cores. We add this to the layer lag. Moreover, due to lag differential between outer and inner layers, the inner layer travels eastward relative to the outer layer; we call this the layer lag.

If we cut Earth at the Equator and view its profile from the North Pole we will see the iso-circular speed curves across the layers consisting roughly of smooth family of counterclockwise spirals that covers the entire profile from the rim of the atmosphere towards the core and winds around it. The speed curves do not cut across the core because of the latter's uniform density and solidity (rigidity); in other words, it spins as a solid rigid core. If we look at each curve there is discontinuity at the interface between the crust and the ocean, between the ocean and the atmosphere and between the crust and atmosphere. Therefore, there is a segment of step function from the crust to the ocean and from the ocean to the atmosphere resulting in a family of piece-wise spiral iso-circular speed curves that covers the entire profile. However, friction rounds off the corners in every curve resulting in a smooth family of curves that approximate step functions. We shall use this information later.

The linear speed of the crust decreases from 1, 609 km/hr at the Equator to 0 at either Pole, *i.e.*, there is a gradual lag in circular speed towards the Poles which are regions of calm being the extremities of the Earth's cylindrical eye.

We shall refer to it as polar lag. Being extremities of the eye there is much suction at the poles which explains the flattened polar regions. Measurement on the Sun shows that midway between the Equator and either Pole the circular speed of the spinning crust is 30% that at the Equator. Thus, circular speed of rotation decreases away from the core's center in all directions.

Energy conservation induces plate movement from region of high kinetic energy which is energy dissipating, such as the region of rapid spin at the Equator, to region of calm at either Pole. Thus, the general movement of the plates in the Northern Hemisphere is northward, in the Southern Hemisphere southward. However, there are particularities that induce slight departure from it. For instance, the huge China Plate and the arrangement of the plates in the region blocks northerly motion; therefore, the Philippine Plate is moving in the north-northeast direction towards the Sea of Japan. The Japanese island of Kyushu southwest of Tokyo, however, which is near and directly south of the mainland has no alternative way of moving north; moves northeastward relative to the mainland which is moving westward as well and is in collision course with it. Such movement and the associated uneven movement of plate boundaries cause earthquakes as subduction occurs and rocks that cross geological faults over them snap beyond their thresholds of tension. Of course, the big ones will come at collision. The sudden movement of huge chunk of crust in the course of such uneven movement also causes earthquake.

Study of fossils reveals that 200 million years ago the land masses belonged to one continent, Pangaea, in the region around the present eastern end of Siberia [107(b)]; 20 million years later the continent began to split into two land masses: Laurasia in the north and Gondwanaland in the south [107(b)]. This splitting is the result of these two main movements: the shearing or slicing effect of the polar lag and the movement away from the Equator, by energy conservation. If we look at the land masses and the seabed we will find spectacular verification of this dynamics. While we can explain the break-up of the single continent Pangaea into Laurasia and Gondwanaland we focus on the later splits and movement of land masses that have direct evidence on the seabed and arrangement of the fragments.

The Ancient continent of Gondwanaland split into India and Antarctica 140 million years ago [107(b)]. India moved northward and collided with the Asian mainland forcing the crust upward to form the Himalayan ranges while Antarctica moved southward and settled over the South Pole. The effect on India was the formation of numerous mountains in the North and along the coasts and plateau at the central region which are non-volcanic, *i.e.*, it was due to uneven compression and crust movement as a result of the collision. The Antarctic is the only land mass connected to the oceans of the globe except the Arctic Ocean, the smallest.

Another example of slicing action by both the polar lag and movement away from the Equator is the Australian Island of Tasmania moving south towards Antarctica. The contours of its northern edge clearly match the Southern coast of Australia where it split from.

A spectacular example of stretching is the North and South American continents cut by the Equator across Ecuador in South America. Clearly, the North American continent moved westward that stretch its link with the South American continent moving eastward that now forms the Central American land bridge. This is an example of a resilient land connection.

Another example of a split where a huge land mass moved westward and a tiny portion stayed put is the land of Sri Lanka that India split from and moved westward. Like Tasmania, the edges of both land masses where the split occurred are still evident.

Still another example of splitting is Madagascar that has been left behind by the movement of the African continent westward; the edges where the split occurred is still evident.

What looks like broken pieces of a land mass north of Canada is also the result of the slicing effect of the polar lag; so is the group of islands north of Siberia.

A huge land mass does not split; instead it moves as a whole. However, if there is a rift it can split or stretch. Fifty million years from now the eastern portion of Africa will split along the Rift Valley and the rest of the continent will move north and close the Mediterranean [107(a)]. The eastern portion will retain the Horn of Africa that includes Somalia and part of Ethiopia as well as Kenya, Tanzania and part of Mozambique [107(a)].

Just recently (*i.e.*, anywhere since 65 million years ago) Greenland split from Europe. These movements are consistent with the combination of the slicing effect on land masses of the polar lag and movement from region of high kinetic energy to relative calm.

These two general movements of the crust are mainly responsible for changes in the crust's landscape including formation of mountains, widening of lakes as well as violent activity such as earthquake and volcanic activity. The latter may even start violent atmospheric turbulence such as hurricane through heating up of ocean surface. In the movement of the Earth's crust, the two parts of Andreas Fault in California are moving in opposite directions that a mountain near Lake Tahoe, California now used to be in Mexico millions of years ago [107(a)]. One can only imagine the earthquakes such movement must have caused. The Red Sea that separates India from Saudi Arabia has expanded during the last 25 million years due to a widening ridge in it and the Red Sea is expected to become wider than the Atlantic Ocean in the future [107(a)]. We can see here that not only land masses change, the oceans change as well. Two hundred million years ago the continent Pangaea was surrounded by the single ocean Panthalassa [107(b)]. Those land masses have spread to their present positions.

The Earth's Magnetic Field

Like its dual in quantum gravity [43] – the atom – a cosmological vortex is a magnet, its magnetic field or gravitational flux powered by its eye, by flux-low-pressure complementarity. In the atom, the protons in the nucleus provide the vortex flux on which the orbital electrons ride [43, 45]; the latter corresponds to the minor vortices of a cosmological vortex. The Earth's initial dark vortex as an eddy in the solar vortex had autonomy. As the dark and visible halo acquired greater concentration along its equatorial plane by centrifugal force, flux-low-pressure complementarity and the dark halo's resonance with the visible halo's dark component, flux compatibility took greater impact on the Earth's and Sun's gravitational fluxes. Then energy conservation moves them towards an equilibrium position: when the Earth's north magnetic pole points towards the Sun or when it points away from it. At present their polar axes are almost parallel with the Earth's polar axis tilted away by a slight angle of 23.45° with the Sun's [85]. They have the same gravitational flux spin so that viewed from their respective North Poles their fluxes spin counterclockwise. This angle is fixed by the Earth's great angular momentum as a huge gyroscope. However, it wobbles (oscillates) across the equator by a maximum of about 25° to the North by June 21 and 25° to the South by December 21 every year. They are called the summer and winter solstices; this wobbling follows from the universality of oscillation and uneven development [85]. As the Earth accumulates mass the oscillation angle will tend to zero asymptotically.

Now, the Earth's magnetic flux reverses polarity at about 25,000-year cycles. Why? Let us learn from the swinging of the clock's pendulum that wobbles about its point of equilibrium. It is understandable since the pendulum has mass and momentum so that when it is pulled away from the point of equilibrium and released it goes back to the equilibrium position and goes past it due to angular momentum; then it swings back towards the equilibrium point due to the pull of gravity and goes past the equilibrium point due to angular momentum, etc. This wobbling or vibration is due to synchronized application of opposite forces, in this case, angular momentum and gravity. In the case of the gravitational flux that is weightless and, therefore, has zero momentum why does it wobble?

We find some clues from the Earth's cosmological history. When it was still dark vortex the gravitational flux was autonomous as an eddy but had fixed orientation. Then as we have already explained it takes an equilibrium position. It is clear from the summer-winter solstices that at the equilibrium position the Earth's polar axis is parallel to the Sun's so that there is neither repulsion nor attraction on either side of the Earth's gravitational flux since the Earth's equatorial plane is normal to the Sun's. We view the Earth from the Sun's equatorial plane with the North Pole to our right.

As the hot spinning core converted dark to visible matter and accumulated mass around the eye, the micro component of turbulence produced seismic waves. Seismic wave is nested fractal sequence of basic cosmic waves that also produces macro visible reverse fractal [36] wave envelope. When the Earth was still light the macro wave induced rapid wobbling about the equilibrium point. As the Earth's inner core gained mass the wobbling became slower but wider. Take a phase of the wobble when the Earth's flux to our right is tilted away from the equilibrium position. Then there is a push by Sun's gravitational flux away from the equilibrium position while on the left there is a pull so that they reinforce each other. As the Earth's gravitational flux crosses beyond the normal angle (where there is neither pull nor push) to the Sun's equatorial plane the push and pull forces switch sides, by flux compatibility, and still reinforce each other but the motion goes on due to angular momentum until it is overcome by flux compatibility and reverses. Due to the effective gravitational pull on the inner core it is carried by this wobble. As its mass grew the angular momentum rose and the wobble became slower and wider, just like a heavy clock pendulum swing, until it reached the present 25,000-year cycle and almost 180-degree of an arc wide wobble.

A familiar example of an infinitesimal vibration that progresses to a wide wobble is the Foucault pendulum. A heavy weight just above ground is attached to a long narrow wire and tide to the ceiling on the fifth floor through a hole on the first, second third and fourth floors. Due to the Earth's rotation the weight vibrates at first in the direction of the Earth's spin. Then, over time, the vibration increases to a wide steady wobble (if above or below the equator the trace is not a line segment but opens up in the middle due to the polar lag).

As the Earth's mass increased its wobble decreased to the summer-winter solstices leaving behind its 25,000-year gravitational flux's cycle determined by its dense inner core that at the same time determines its magnetic field to this day. The summer-winter solstices developed independently of the inner-outer core's wobbling since the mantle and crust have independent spin due to lesser density, greater flux lag and, hence, less pull by the gravitational flux beyond the inner-outer core. The increasing mass of the mantle and crust have breaking effect because the gravitational flux has more effective pull only at a narrow angle with the equatorial plane. Therefore, the wobble loses energy during the 25,000-year cycle but it is replenished somewhat every 25,000 years when the gravitational flux plane is within that narrow angle of angle of 23.45° with the Sun's equatorial plane. That is the situation now. We can then expect that the summer and Winter Solstices will grind to a halt in the future and the Earth will become a huge steady gyroscope even if the inner-outer core also continues to wobble along with the gravitational flux. Just as the Earth's gravitational flux determines its magnetic flux, so does the latter determines its magnetic flux's polarity and its 25,000-year cycle.

Both the magnetic field and summer and winter solstices wobble will stop before the Earth's vortex becomes isolated around the eye and the core reaches its destiny – black hole. The magnetic field wobble, radial oscillation that leads to elliptical planetary orbit and summer and winter solstices belong to one category: they all follow from the universality of oscillation and law of uneven development.

The Earth's Trek Back Home

The eye of a cosmological vortex is a region of calm and depression (low pressure) like the eye of any vortex, e.g., typhoon and tornado, except that it contains nothing but great concentration of non-agitated superstrings called black hole. Therefore, it de-agitates visible matter at its boundary (event horizon) each de-agitated superstring joining the black hole in it and, over a very long period of time, converts them to semi-agitated superstrings first then to non-agitated superstrings that accumulate and join the black hole in the eye.

For comparison, let us look at a star. By energy conservation, over a very long period of time, the kinetic energy of its interior including its angular momentum and molecular bonding weaken. Then its component prima split into separate prima held together only by the suction of the eye. This phase of its life cycle is called neutron star, a misnomer since there is no neutron there. It is a transition phase where the prima eventually collapse to semi-agitated superstrings, by energy conservation, then to non-agitated superstrings that accumulate and join the black hole in the eye. Being dark, the black hole is unaffected by centrifugal force and remains in the eye's center. When the prima have significantly converted to non-agitated superstrings and joined the black hole then it becomes naked. Naked black holes in the Cosmos are in pockets of "non-activity" their surroundings appearing empty. Many of them have been catalogued by astronomers. Clearly, black hole does not suck; only the eye of the cosmological vortex that nurtures it does. The Earth follows the cosmological path of a star towards its destiny – a black hole – at much smaller scale.

It is clear that every cosmological vortex with minor vortices including our universe follows this evolutionary path to its destiny, a cluster of black holes back in dark matter. The only minor vortex of the Earth is the moon; its destiny is also a black hole.

Clarification of Issues and Natural Phenomena

We clarify some issues and natural phenomena from the perspective of GUT.

Stability of our Universe

Traditional cosmology presents an unstable universe of ours by assuming the anthropic principle that says essentially that if events in our universe did not occur exactly as they did our universe would have collapsed [74]. This would have occurred, for example, if at some point in its formation there was one proton less. GUT's qualitative model of our universe presents a stable universe that upholds energy conservation from the Big Bang through its destiny in dark matter. Moreover, dark matter is stable and serves as absolute frame of reference; it resolves Einstein's twin paradox.

A Paradox no More

Originally referring to our universe, Olber's paradox [74] says that it cannot be infinite; otherwise, accumulated light coming from all directions would have fried us in intense light by now. Now moot since our universe is finite we broaden it: can visible matter be infinite? There is no reason to rule this out. There are two possibilities: (a) visible matter is suitably dispersed that light reaching our universe does not accumulate and dissipates energy before reaching us or (b) light reaching the vicinity of our universe is deflected away by its gravitational flux.

Transitory Natural Laws

Natural laws are revealed by natural phenomena. Before the Big Bang there was none; after the Big Bang, natural phenomena appeared with increasing complexity that revealed new natural laws. Now, new natural phenomena exist, e.g., biological phenomena, that reveal biological laws. They will gradually vanish as our universe takes its trek home to its destiny.

Point of no Return

There are three speculations about our universe (a) steady state, that it will remain as it is forever, (b) pulsating universe, that it will eventually reverse its present expansion and head toward a big crunch that can be viewed as another big bang and start a new universe and (c) forever expanding universe.

Obviously (a) does not describe the present state of our universe which is steadily expanding at accelerated rate. Item (b) violates energy conservation; it would require staggering amount of force to reverse the huge momentum created by 8 billion years of acceleration due to its centrifugal force of spin and there is none of it over the horizon. The only other force in the direction of reversal is gravitational suction which can only diminish with the thinning of dark halo. Furthermore, there is no evidence among the billions of galaxies that it can happen. The most that can happen is weakening of the gravitational flux in which case the galaxies will be on their own heading to their destiny. This is exemplified at a smaller scale by the status of the Sagittarius cloud of stars [82]. With respect to (c) while galactic momentum may sustain the galaxies' outward flight dark viscosity will catch up with and constrain them to a halt as they trek back to their destinies. At the same time, minor vortices still in orbit and others at the periphery now freed from gravity will steadily fall and join the collected masses around their eyes, enter the transitional phase of de-agitation and ultimately join the black holes in their graveyards, their eyes. This means that (c) is impossible. What will happen then? Our universe will continue to expand, reach its peak and the core will lose influence on the minor vortices as dark halo thins out so that the minor vortices will be left on their own, each vortex will approach its graveyard, its eye, and join the black hole there.

Supernova

The current understanding of supernova as explosion of star, that it is part of the evolution of a star, is incorrect since cosmological vortices are stable. Moreover, a supernova is a rare phenomenon and does not quite match the fact that there are trillions upon trillions of stars in the Cosmos. Left alone, a star evolves towards higher order, a black hole in its eye. Therefore, the only plausible explanation is collision of two stars of opposite spins that approach each other along their common equatorial plane. If they have the same spin they avoid each other, by flux compatibility. With opposite spins they attract and, by their momentum, the fluxes between their eyes merge smoothly at first until the rim of one goes past the eye of the other and their fluxes, being opposite, collide resulting in double explosion. Then the flux barrier between their eyes breaks and creates huge depression that violently sucks matter around causing more powerful third explosion. This phenomenon is dual to primum-anti-primum mutual destruction of quantum gravity. (Photographs of supernova show the three rings of visible matter on expanding shock waves corresponding to the three explosions [108]).

Background Radiation

It is believed that background radiation verified by the COBE project [106] is a relic of the Big Bang. Whatever happened during the Big Bang was obliterated by ensuing events, e.g., Cosmic Burst [37]. Even our universe's present accelerated expansion has nothing to do with it but is due to our universe's centrifugal force of spin as a super…super galaxy that throws the galaxies outward at staggering speed [98]. All cosmic waves including the so-called background radiation are generated by atomic nuclear vibration [42], micro component of turbulence at the cores of cosmological vortices, explosions in the Cosmos such as supernova and possibly black hole explosion [102].

Physical Dimension

While we have a good sense of mathematical dimensions there is no theory of physical dimension at this time. Therefore, we borrow some ideas from mathematics to devise a framework for a physical theory of dimensions.

Concluding Remarks

We have seen the relativity of relevant dimensions depending on the subject matter or discipline of study. For physics, the basic frame of reference is the x-, y-, z-coordinate system and, depending on the area of study, more dimensions may be added. Such dimensions are determined by the relevant independent motions impacting on the objects of study. For the study of dark matter independent motion is due to the impact of cosmic waves coming from all directions in the Cosmos. However, they have dark frequencies (the reciprocal of dark wavelength which is less than 10^{-14} meters) that resonate only with or have impact on objects of comparable order of magnitude. For instance, for the study of motion of ordinary objects up to atomic scale, three dimensions are sufficient since cosmic waves resonate with their dark component and have no visible impact. However, for the study of the structure and motion of parts of the atom another dimension must be added because cosmic waves resonate with and vibrate their dark components and, hence, the atom itself.

Why one dimension when cosmic waves come from all directions and why only simple vibration? Would not the impact be more complicated than simple vibration which is one dimensional? It follows from the Internal-External Dichotomy law of nature that determines the reaction of those objects, in this case, vibration. Then the atom itself is also vibrated. Perhaps, we can answer this question using an analogy with the tuning fork. If we strike it simultaneously from different directions it will only vibrate with the same characteristics and, hence, the same pitch and tone, determined by the way it was manufactured, *i.e.*, its internal structure. Those studying the superstring need 11 to 26 dimensions. Theoretically, the possible number of dimensions is a large number because the basic cosmic waves alone are generated by the atomic nuclei which are everywhere in the Cosmos and are highly energetic waves that travel at great distances measured in light years. Incidentally, Lorentz transformation of the theory of relativity reduces the dimension of the space from four to three with the dependence of time on one of the coordinate axes.

Finally, the grand unified theory is indeed unified in this sense: the superstrings comprise dark and visible matter, semi- and non-agitated superstrings comprise the former and non-agitated superstrings comprise the latter.

CHAPTER 4

Applications

Abstract: GUT and its methodology (qualitative modeling) opens up new fields in physical psychology, biology, geological and atmospheric sciences, oceanography and medicine, and brings the science community to the threshold of the new epoch of GUT Technology. The summary of the solution of the gravitational n-body problem brings us to a full circle since the journey into theoretical physics, cosmology, philosophy and applications were started with this problem.

Its theoretical application to physical psychology is the underlying physics of the mind and here are its findings: (a) brain waves are the common medium for carrying out the brain's and gene's functions; (b) the brain controls entire body functions and processes, its principal function, and the cortex, the thin layer that wraps around the upper lobe of the brain, is the center of thought, its secondary function; (c) the main function of the gene is to produce body tissues in the cellular membrane and ensure that every part of the body grows in the right place. (d) the principal factor in evolution is the gene.

The theory of chaos and turbulence is developed in the geological and atmospheric sciences and applications, while wave motion and water current are explained in oceanography. The theory of genetic composition is developed towards the generation of technology for the treatment of genetic diseases.

An overview of the technological applications of GUT is provided and the common thread is the utilization of the clean, free and inexhaustible dark matter that is abundant everywhere in the Cosmos.

INTRODUCTION

Our applications of GUT are both theoretical and practical. The former refers to development of physical theory, *i.e.*, qualitative model of reality in natural science including biology [28, 30, 31], genetics [30], geological and atmospheric sciences and oceanography [35, 38], physical psychology [28, 29] and medicine [32]. Practical applications refer to solution of scientific problems and development of technology that runs on conversion of dark or latent energy to visible or kinetic energy called GUT technology and applications of GUT to medicine such as treatment of genetic diseases, e.g., cancer, systemic lupos erythematosus, diabetes, muscular dystrophy and mental disorder, e.g., depression, without injury to normal cells.

Since physical theory explains natural phenomenon it is, in a sense, verification of prediction of an occurrence in reverse because the natural phenomenon occurred in the past and continuous to evolve to this day. This is what GUT does in many of its applications including devising technology which involves some prediction.

THE THEORY OF INTELLIGENCE

We identify the domain of the theory of intelligence. Any activity that occurs in the cortex of the brain (thin layer that wraps around its upper lobe about ¼ inches thick, convoluted and has effective surface area of 16 square feet) is a component of intelligence. It includes sensation and memory; reflex belongs to the secondary nervous system centered in the spinal column and extending to the system of nerves. The combination of intelligence with emotions and feelings centered in the heart is what we call the mind. Intelligence includes instinct, a notch higher than reflex and peculiar to higher forms of animal species including reptiles and primates. For instance, it requires some intelligence for the (blind) snake to assess the report of its sensory organ, the tongue, and catch a prey. There are certain activities of plants and animals that lie between instinct and reflex. For example, through genetic alteration during its evolution a species of pitcher plant developed a contraption for catching mice for its food; so does the string bean that competes with the bitter melon over a trellis to climb on. The string bean wraps around the bitter melon and kills it.

Over time, the bitter melon has learned its lesson. When planted near a string bean it veers away from it. We consider instinct a component of intelligence being a combination of memory and experience.

Intelligence is the brain's secondary function; its primary functions: control and coordination of body processes and movements.

The Human Brain

The human brain weighs 2.2 pounds. On top of the upper lobe and around it above the ears is the cortex. Its neurons have dendrites and axons sticking out of their nuclei. A sequence of brain waves carrying signals ("information") lights up the tips of two dendrites like radio transmitter antenna as it jumps across the dendrite tips from one neuron to another and establishes communication channel by vibrating and activating them and letting through signals with the same characteristics (resonance) as the previous wave that passes through. Signal of different characteristics opens up new neural channel.

From behind the forehead through the top of the cortex is the creative-integrative region (CIR); around it on the left, back and right of the upper lobe are the sensation regions. A concept is modeled in the CIR as suitably activated network of neurons. They are composed by brain waves carrying signals, the neurons vibrated via resonance. During sleep the concept's components are dispersed and encoded in the appropriate sensation regions (sight, hearing, taste, smell and touch) as corresponding suitably vibrated neural network for storage. By energy conservation, a concept's neural network in the CIR is de-activated as its components disperse to the appropriate sensation regions. When incoming signals stop the activated neural network reverts back to normal vibration. The CIR recalls and recomposes them into the same neural network of the same concept if needed for mental activity. Recall of a concept occurs by direct cosmic agitation of its sensation components or the CIR directing and focusing cosmic waves to them. Then the component neural networks in the sensation regions are de-agitated, energy conservation, as its component vibration networks are recalled to and re-composed in the CIR.

Psychologists know what happens in the brain but not the physical processes involved, e.g., during sleep what is learned is transferred to some part of the brain. Now we know that the information learned is transferred to the appropriate sensation regions through the formation of the component neural networks that model it.

The brain is the most active organ of the body. It uses 1/4 of the energy from food most of which spent on mental activity. Mental activity includes concentration and directing cosmic agitation to the sensation regions in search of concepts components for recall to the CIR to draw conclusions or make inferences and commands to execute conscious activity and compose new concepts. They involve transfer of energy from agitated to non-agitated neural regions subject to energy conservation.

Brain Waves

Cosmic waves are the prime mover of our universe. Every piece of matter vibrates due to the impact of cosmic waves from all directions, its vibration characteristics principally determined by its internal physical structure [42], e.g., molecular structure, induced on nuclear vibration characteristics of the nucleus. Every piece of matter has temperature above absolute zero due to its motion; semi- and non-agitated superstrings have absolute zero temperature. Moreover, thought is triggered by cosmic waves and consciousness is due to the normal vibration of the neural network in the CIR. For our purposes here, we focus on basic cosmic waves. Since the nucleus is fractal, so are basic cosmic waves which are endowed with huge energy. When basic cosmic wave is encoded with the vibration characteristics of a living cell it is called brain wave. Being highly energetic electromagnetic waves can penetrate great barriers, e.g., one can tune in the radio in a closed room. Encoding by neural vibration does not diminish the energy of electromagnetic waves.

Resonance

An important physical principle governing wave interaction is the principle of resonance that says: maximum resonance between waves or vibrations occurs when they have the same wave characteristics but its principal determinant is frequency or wave length [40]. In practical life resonance regulates radio and TV reception by frequency. When a tuning fork of certain frequency is struck its vibration resonates with and vibrates a tuning fork of the same frequency nearby to produce the same pitch. When two waves resonate, *i.e.*, they have the same order of magnitude of wavelength they produce composite wave with modified characteristics.

When resonance between waves or vibrations is disruptive it is interference or discordant resonance. Observation of an object in the Cosmos (using light) is possible when the object bounces light that hits it or silhouetted against a source of light or causes interference that occurs when its size is at least the same order of magnitude as the finest wavelength of visible light, 10^{-14} meters, since at least some wavelength of light is distorted or demolished and detected by the spectroscope. Objects of lesser order of magnitude, e.g., semi- or non-agitated superstring that comprises dark matter, are not observable directly, its existence verified indirectly by the impact on visible matter in accordance with energy conservation. Resonance between waves reinforces each other because their energies add up just as fluxes of the same direction do when their speeds are within the same order of magnitude.

Sensation and Concepts

Any sense organ has two components, receptor of visible signals and convertor to brain waves for transmission to the CIR through the neural cells. Some sense organs have the same receptor and convertor, others separate.

Consider the organ of taste, the tongue; its receptor consists of separate groups of taste buds (for sweet, sour, bitter and salty tastes) along its edge differentiated by visible vibration characteristics that make them resonate with visible waves generated by food molecules of corresponding tastes. The most accurate medium for molecular vibration is liquid, saliva in this case. Take sugar molecules. They must be soaked in saliva so that molecular vibration is transmitted accurately and the taste buds for sweetness, attached to the nerves at base, resonate with and are vibrated by it. Their nuclear vibration wiggles the basic cosmic waves they generate, superpose or encode the vibration characteristics for sweetness on the normal nuclear vibration that generates basic cosmic waves and turn them into brain waves which are transmitted to the CIR through bundles of linear chains of nerve cells.

Understanding taste sensation has yielded technology for control of diabetes. Insulin brings in sugar to the cells for nourishment. When pancreatic secretion of it is inadequate, a condition suffered by diabetics, blood sugar level rises. Sugar is viscous and raises high blood pressure that can damage fine blood vessels of the kidney and eye and cause other complications. Since taste is due to the physical structure of food molecule that determines its vibration characteristics, why not simulate it using non-viscose materials? In fact, this is what synthetic food is. Synthetic sugar "Equal" is suitably designed carbon molecules with the vibration characteristics of sugar molecules; it is 550% sweeter than ordinary sugar but without harmful viscosity [28, 29, 30]. Now there is great number of synthetic foods, e.g., beef and crab "meat" from vegetables, with the taste but not the cholesterol.

As diverse characteristics carried by brain waves through single linear chain of nerve cells reach CIR each characteristic connects and activates (vibrates) a linear chain of neurons so that the various characteristics induce formation of linear chains of neurons joined at a common nodal point in the CIR. The length (number of neurons) of a branch of activated linear chain is proportional to the duration of that characteristic of the event. The number of distinct characteristics is equal to the number of branches of linear neural chains at this nodal point. The intensity of the component of an event is proportional to the size of bundle of linear chains that carry the characteristics to the CIR and its duration is proportional to the respective lengths of linear chains of neural cells it encodes and activates. The union of nodal regions of the neural network of various components of an event is the nodal region of its induced full concept. When the same event is repeated it resonates with that same neural network. Same event of greater intensity adds more linear chains to its bundle.

However, another factor enhances intensity of signals: neural cell ionization of sodium atom that creates electrical potential followed by its jump to lower potential accompanied by electrical impulse. What causes this ionization, *i.e.,* expulsion of orbital electron from the atom? The only force that does it comes from the energetic brain wave passing through. The sodium jump amplifies nerve cell vibration and, hence, intensity of signals encoded in brain waves.

The entire neural network activated by an event reverts back to its normal vibration as soon as it is over and the concept induced by the event is now encoded. During time of activation and encoding sensation dendrite tips spark and propagate brain waves across dark matter. Energy conservation requires that when encoded neural network is deactivated its components are transferred to and activate similar neural network in corresponding sensation regions. As the transfer is completed the network reverts back to normal vibration and the components are now stored as long-term memory. This transfer usually occurs during sleep.

When a sensation region is hit by suitable basic cosmic waves directly from the Cosmos or indirectly through the CIR the encoded neural network there vibrates with exactly the same characteristics as when it was activated and encoded. The vibration recalls some concept component to and vibrates its nodal region in the CIR. In turn, it vibrates and recalls other components and recomposes them there. Then the individual recalls the concept and dendrite tips light up and propagate brain waves. Thus, the psychologist's prescription for recalling non-physical concepts by association is correct. For instance, the concept of time is the result of the CIR's creative capability; it is neither an event in the real world nor associated with sensation nor does it have physical referent. Therefore, time is determined only in association with contiguous events.

How does one recall the year a song became a hit? This requires association by the CIR with the right contiguous events that induce physical concepts such as one's place of work, friends and associates, etc. Their recollection (activation of neural network) fixes the year.

Agitation of concept's nodal region vibrates its neural network in the sensation regions and recalls its components to the CIR where they are recomposed around it. An event may have several components. For example, a fire cracker explosion has light, sound, smell and touch components. After an event its components are sent to their respective sensation regions through the CIR automatically, by energy conservation, but available for recall by the CIR.

When the nodal region is activated directly by cosmic waves, re-composition and recollection is spontaneous and simultaneous with sparks and propagation of cosmic waves at dendrite tips. When concepts are recalled, dendrite tips of their neural network also spark and send out brain waves with vibration characteristics corresponding to the information they are encoded with. Since mental activity involves passage of vibration characteristics, it lights up dendrite tips and propagates brain waves. Therefore, recall of concept components to the nodal point in the CIR or dispersal to the respective sensation regions for storage involves passage of energy from activated to non-activated network and quite analogous to passage of gas or liquid from high to low pressure. While generation and propagation of brain waves are well understood and explained by natural laws reception is not fully understood. One such unexplained phenomenon is mental telepathy (ESP) where the CIR receives information from external source directly. There is no systematic study of this phenomenon on humans but there is scientific confirmation in dogs.

Recall that in taste the taste bud is both receptor and convertor; not so with hearing. The eardrum is receptor of ordinary sound waves. The outer ear catches and directs sound waves through the outer ear canal to the eardrum which is vibrated by resonance. On the other side of the eardrum is the inner ear that contains the cochlea, a spiral canal filled with liquid. Near the eardrum immersed in the fluid are the three tiniest bones in the human body: malleus (hammer), incus (anvil) and stapes (stirrups), in triangular arrangement that detects the direction the sound comes from. Lining the interior of cochlea are strands of hair separated according to vibration characteristics. They are the convertor of sound waves to brain waves for transmission to CIR and composition as concepts. Sound waves from outside vibrates the eardrum where the handle of the malleus is attached. As the eardrum vibrates, the hammer hits and vibrates the incus. The stapes stirs the cochleal fluid, amplifies vibration and propagates it, the latter resonating with and vibrating the corresponding hair strands that line the inner wall of the cochlea. Their atomic nuclei vibration encodes the characteristics on basic cosmic waves (now, brain waves) that carry and transmit the information to the CIR that converts it to neural networks. Thus, the hair strands are convertor in this case. In due course its components are encoded and stored in the appropriate sensation regions. Energy conservation insures that when concept's components are stored in the sensation regions its neural network in the CIR deactivates and restores normal vibration and vice versa.

By energy conservation, the energy of brain waves carrying component characteristic of an event is proportional to its duration, hence, length of linear neural chain it activates; its intensity is proportional to the number of such neural chains (bundle). Moreover, by energy conservation, the total energy expended during an event is proportional to the length of the neural bundle chains it joins, activates and encodes. Thus, the more powerful the event the broader the neural network it composes, activates and encodes.

Clearly, every concept has nodal region that joins all its components together. When agitated by suitable cosmic waves it vibrates, recalls its components and recomposes them around its nodal region. When one component is agitated the nodal point is agitated and, in turn, vibrates and recalls other components and recomposes them in the CIR.

There is one more missing link on sensation: how does feeble encoded vibration of receptors get transmitted to the CIR and corresponding sensation regions? Neural bundle from a sense organ through the sensation region has the same vibration frequency and the CIR has all the frequencies of the neural bundles from the sense organs. Therefore, the neural networks in the CIR and sensation region resonate with and are composed by brain waves from corresponding sense organs by resonance. As encoded brain waves from a sense organ pass through the nerve bundles they ionize and raise concentration of positive ions, e.g., sodium, in the neural membrane creating electrical potential between membrane and cell interior. It is resolved by sodium jump that vibrates the cells and produces electrical impulses at corresponding frequencies as those at the convertor that establish resonance and replicate and reinforce vibrations all through the CIR and corresponding sensation regions.

Value, Perception and Cognition

The CIR can recall and recompose concepts components at their respective nodal regions.Psychology says 80% of one's perception of event comes from memory, only 20% from outside. The psychologist uses this information to assess one's personality by asking the subject to tell what he perceives from a set of objects. It reveals the contours of one's life experience recorded in the sensation region.

Values and one's way of drawing out conclusions or logic that we include among values are learned in accordance with Pavlov's theory of learning or conditioning. Values consist of a system of "correct" choices, decisions, conclusions, etc., corresponding to the right neural connections to their respective concepts in the CIR. They are formed through training, peer pressure and experience. This is the basis of putting one's logic or way of thinking in the category of *value*. Correct choices are imbibed by approval and incorrect ones by rejection first by the parents and, later, by teachers and peers. More values may be learned from experience and training. Correct choices, decisions, etc., are modeled as encoded as neural connections to their premises by neural network in the CIR. A discipline of knowledge (e.g., mathematics or physical or social science) consists of system of concepts and values encoded as suitable neural network whose nodal regions are connected to the central neural region of the discipline in the CIR at which its axioms or natural laws (of physical theory) or social principles (of social science) are recomposed. Different disciplines of knowledge may be autonomous but connected through their central nodal regions so that the person can switch focus from one discipline to another through their respective central neural region.

However, for coherence, they are connected to the central neural region of the entire knowledge of the individual. In rare cases distinct and disconnected systems of values are involved and the individual suffers from multi-personality problem where one personality may not know the others. A normal individual may have several connected systems of knowledge including practical knowledge that allows him to function day-to-day. Epilepsy is a condition where there is anomaly in coherence of an individual knowledge system.

One's thought is modeled by physical processes and neural network in the cortex, its representation as language is in the real world. Scientific knowledge articulated as qualitative mathematics expressed in suitable language is built, stored and applied as physical theory anchored on some laws of nature. That is how thought and knowledge are related. However, as we have noted previously this representation is ambiguous so that knowledge is built on the unambiguous representation of thought as knowledge. Thought allows the individual participation in building that knowledge.

During, say, a game of bowling the player models the lanes at the central nodal region by appropriate neural network where the nodal region of his lane is the focus. The CIR commands the body to play the game on the basis of its perception of the lane, knowledge of the game and previous experience. A skillful bowler accurately models the lane on the basis of which the CIR commands execution of the game.

In any physical endeavor thought gives command through brain waves that activate appropriate parts of the body to execute it. Sometimes thought triggers automatic execution. For instance, when one thinks of lemons brain waves convert superstrings to saliva in the salivary gland. Thought can now move the PC curser and hit an icon or even operate the motor of prosthetic arm.

In theoretical research the mind utilizes his knowledge, observation, experimental data, training, social or physical theory, if any, and mathematical tools to "discover" and articulate (create) laws of nature upon which to build a theory. Although natural laws refer to physical concepts their statements are mathematical principles because they have no physical referents, *i.e.*, one does not find them in the external world. They are created and articulated by the mind in the appropriate language based on synthesis or analysis of known information, observation and experimental data.

For the other function of the brain, as control center of all processes in the body, stress dulls its ability to keep the auto-immune system in top shape that the latter is unable to do its normal function of destroying unwanted systems in the body. Ordinarily, the body has normal level of cancer cells and the auto-immune system adequately destroys and keeps them at tolerable level. However, under stress, it may fail and be overcome by their replication; then the person becomes sick. That is why healthy body and mental health go together. In some cases, extreme stress or emotional trauma may cause physiological imbalance such as raised level of brain regulators, e.g., serotonin. This is what happens in depression where the neural network vibrates even when they are not supposed to and the individual hears voices or sees things (audio or visual hallucination) that are non-existent, or makes unusual decision, e.g., attempts suicide. Such imbalance may be due to a gene or emotional trauma; in the latter physiological change may occur that is eventually encoded genetically in accordance with the law of genetic encoding [30, 31]. In general, strong emotion is accompanied by physiological changes, e.g., raised adrenalin secretion associated with anger. Strong sensation may also cause physiological changes, e.g., extreme pain produces molecule in the neural membrane [76, 84].

When the CIR lacks ability to direct or control cosmic waves the person suffers from a condition called autism where he cannot concentrate or focus. In one form the person is simultaneously aware of everything around him and cannot respond properly. Naturally, he has short attention span, since diverse sensations compete for attention, and becomes hyperactive (suitable therapy can now control this condition).

Learning and Creativity

We broaden our discussion to the mind since feeling which is in the heart is also learned. However, creativity is specific to thought.

Learning is fitting new concepts, information, etc., into knowledge already in the mind. A normal mind is both selective and focused. It entertains only very few signals from the external world. For example, most students are aware of what happens in the classroom but unaware of the noise outside. Of those registered input, only a few is retained as long-term memory, *i.e.* the components are encoded in the appropriate sensation regions. What is the basis of long-term memory? It is resonance in the broad sense. New information is learned or retained when it fits a knowledge system in the mind based on logic or system of values particular to it. This process involves neural encoding and connection to appropriate nodal region of this knowledge system. It is active mental activity that requires concentration.

Moreover, information that requires deep concentration to understand and integrate into a knowledge system is never forgotten once grasped. Some students prefer to listen to and concentrate on the lecture rather than take notes, which can be distracting. This way they grasp and learn the subject matter more effectively.

Another source of information is creativity. The mind can compose new concepts that fit into the individual's knowledge or compose new set of principles or even do deductive or inductive reasoning based on existing knowledge system.

What facilitates learning? One is critical thinking, a core value of science and mathematics aside from creativity. In fact, creativity is the offspring of critical thinking. Taking a critical look at existing knowledge the mind identifies weakness that triggers search for remedy. It was this critical attitude that triggered critique-rectification of foundations and the real number system that paved the way for the resolution of Fermat's last theorem. Moreover, critical thinking increases mental activity the effect of which is to deepen grasp of the subject matter, *i.e.*, enhance neural network interconnections, and induce long-term retention and retrieval.

The critique of foundations of physics, particularly, its methodology undertaken in 1997 [41] revealed inadequacy of quantitative modeling. The remedy was qualitative modeling that gave rise to theoretical physics. We shall identify learning principles appropriate for math-science education and theoretical research.

Learning Principles

The learning principles here are synthesized from theory and experience.

Principle of Complexity and Retention

Learning effectiveness increases with the level of complexity of the subject matter and mental processing required; the more complex the subject matter the higher the level of concentration required to grasp it and the more permanent the retention.

One's ability to concentrate is enhanced through concentration in the midst of great distraction; this is analogous to the greater grasp of subject matter the more complex it is. The rate of learning graph as function of time approximates the logarithmic curve that tapers off away from the origin. This is familiar to psychologists. The tapering reflects the law of diminishing return as the mind gets tired, loses interest and concentration drops to zero. When learning requires high level concentration due to the complexity of the subject matter the rate of retention graph rises towards a vertical asymptote at the completion of the learning process signifying permanent retention or memory. This is true of complicated scientific analysis or principle. When learning involves only mechanical activity, e.g., card concentration game, the curve drops off to zero when the activity stops.

Scientific research involves creation of new concepts, principles and their integration into mathematical or physical theories. As language, mathematical principles are man-made, borne out of extensive experience and study. Physical principles are also man-made but inspired by or drawn out of observation and suitably articulated by the scientist.

Learning and Resonance Principle and Recollection

The mind only imbibes information that makes sense, i.e., fits some knowledge in it and, in mathematics and science, processes only information that fits some mathematical or physical theory.

There are, of course exceptions, like trivia, where there is motivation to win some prize at TV quiz. Retention is also enhanced by the diversity of sensations involved in learning something.

Learning is not a creative process unless it comes from original research. Creativity, e.g., obtaining results from frontier research, comes from extensive experience and knowledge and condition of the field of research. For instance, research creativity is limited in fields already well explored. This means that mainstream research is not necessarily productive.

Retrieval or recollection is not learning but looking for information already encoded in some neural network in the sensation regions. However, it can increase one's receptiveness to related incoming information and enhance learning resonance. In some cases the incoming information may trigger recollection.

Absent-mindedness is due to weak concentration at time of encoding. For instance, one picks up an umbrella as a reflex action to rain; he is likely to leave it somewhere when the rain stops.

The material basis for recollection is agitation of the neural network that stores the components of the information. Then they are recalled and recomposed at their nodal regions. Activation of a sensation component of a concept also activates the other components through their common nodal region. Thus, it is difficult to recall only the color of or half a ball without recalling the whole piece. This confirms that sensation occurs in the CIR. Contiguous events are encoded neural network with common interconnected nodal regions in the CIR.

Well-grasped complicated topics are easy to retrieve due to enormity of concepts and, therefore, interconnected network involved. Activation of one network triggers activation of others connected to it.

Dreams are due to actual neural activity triggered by cosmic waves. They are often incongruous because the conscious function of the CIR – focusing and directing cosmic waves and recomposing concepts – are missing during sleep resulting in incomplete activation of conceptual components or incorrect re-composition of concepts.

However, the content of a dream is culture based determined by encoded neural network in the various sensation regions.

Principle of Association and Neural Network Interconnection

The more interconnected and contiguous the encoded neural network the easier and more precise is the recollection.

The mind learns not only through cognition of similarity and regularity of patterns but also through comparison and contrast by drawing certain conclusions at extreme situation and dramatizing the impact of opposite conclusions. Stressing a point and reducing an undesirable conclusion to absurdity are among the practical tactics for effective teaching. Stressing a point broadens the neural bundle activation involved and enhances the permanency of retention. Moreover, reduction to absurdity or contradiction jolts the mind into recognition of error and negation of a particular point. They raise the level of interconnection and enhance the ease of recollection and clarity with details. Conversely, concepts not recalled for a long time are likely to be drawn into the subconscious. Physically, this means that when the associated neural network remains un-agitated for a long time, it becomes difficult to agitate and retrieve the information encoded in it. This is analogous to the phenomenon of atrophy. Psychologists have devised techniques for drawing out information from the subconscious or, perhaps, blocking off unpleasant memory. In the former it amounts to increasing agitation of the neural networks of contiguous events around the information being recalled and, in effect, raising the intensity of agitation. In the latter it amounts to altering the recollection of previous events. In some cases, suitable psychological technique can make an innocent person confess to a crime. Constant repetition of a lie may cause it to be perceived as truth. This technique is often used by protagonists at war.

Related concepts have connected neural network. When the brain is engaged in certain activity one or more of the nodal regions are activated. This results in activation of what might be called boundary conditions through which the nodal regions of decisions, judgment or commands are activated.

Structure and Coordination of Knowledge

A discipline of knowledge is encoded as system of neural networks with interconnected nodal regions all connected to a central neural region. Different disciplines have interconnected central neural regions, each autonomous but activated one at a time or in specific groups due to the CIR's ability to concentrate and direct activation.

When the mind is engaged in research, e.g., physics, the neural networks of related disciplines, e.g., mathematics, also activate when needed. This means that there is some super nodal region to which all the nodal regions of the various disciplines are hooked up.

Some parts of experience are recorded in the subconscious, *i.e.*, network not vibrated for a long time. Ultimately, neural physical characteristics are genetically encoded and, when the neurons die, they are passed on to their replacement. Only physical characteristics, not concepts or knowledge that are encoded only in neural networks, are genetically encoded. An example of such physical characteristic is the cosmic intensity threshold to activate a network. Another is the encoding sensitivity of the neurons. Persons with photographic memory have high encoding sensitivity and low degradation of neural vibration characteristics. This information might have implications for the study of intelligence, particularly, learning capability.

However, there are rare cases where neural network interconnection is genetically encoded, not just the physical characteristics of the neurons. An example is Devi Shakuntala's ability to compute huge numbers faster than the computer and yield the result quickly without her really knowing how she did it. The only possible explanation is that there was in her cortex suitable interconnected neural network that activated and carried out the computation automatically as soon as the numbers were read to her in the same way reflex is activated by certain stimuli. Both this special skill and reflex are genetic. After all, the synapses that join the dendrites are cells that have genes that determine their physical characteristics. Such exceptional ability is likely to arise in old cultures due to genetic and evolutionary factors. Devi Shakuntala was an Indian and India has the oldest culture dating back to the Vedic Civilization 25,000 years ago [79]. Records suggest that the subcontinent went through the ice age twice (the ice age occurs at interval of 10,000 years) [79].

Creativity is not a physical characteristic but a mental capability resulting from training and extensive experience and one's knowledge, especially, with regards to nature. If the physical characteristics of the neural cells are good, *i.e.*, they are good genetic material, mental capability is enhanced. Instinct is also genetic. For instance, it is in the gene of the female lion to be a good hunter but the female cub needs training by the mother to actualize it. The mother borrows the prey's cub early in the day for the cubs to practice on and returns it later.

Mental activity is indirectly observed as electromagnetic sparks. It has been modeled quantitatively by differential equations [88, 89, 90].

As noted already, the mind is highly selective since neural network responds to and registers only signals having characteristics similar to what is already encoded in thought, a form of resonance. Interactions occur in accordance with the resonance principle [31]. This is true of light; it can 'spot' only objects of order of magnitude in size comparable to its wavelength; in social matters an individual does not blend well in a group with different system of values. The mind is also selective via resonance or conformity with its values. Logic in the mind is altered only through critique-rectification; in fact, formal logic has been upgraded through this technique [21, 78].

Learning-Efficiency-Memory Principle

Learning efficiency is in direct proportion to level of memory encoded in the cortex.

This is confirmed by the fact that about 80% of what we perceive is drawn mainly from memory. The mind fills in gaps in memory and makes sense of incomplete information to trigger mental processing. Much of what we see, for instance, comes from the sensation regions but integrated with external signals and composed by the CIR.

Memory is pre-condition for mental activity and involves two phases: encoding and retrieval. Permanency of retention rises with level of concentration and receptivity of neural network combined with intensity and diversity of sensual components that fit some system of values. Diversity expands the neural network activation and enhances recollection. Intensity enhances details. The signals may come from sense organs as specific characteristics of brain waves.

Diversity of signal components also enhances retention, e.g., when the signals are coming from different sense organs and network. Writing a concept involves sight and touch on the part of the person doing it. In the classroom the student has the benefit of both sight and sound. One can try this experiment: write the items you want from the grocery store and intentionally leave the list behind. Then you will be able to buy more of the items than if you just thought of them in the store. The reason is: when the grocery items are listed at least three sensations are involved: sight, touch and signals from other neural network (thought). This is the reason for the telephone touchtone: the sound pattern enhances retention. For some, once he dials a phone number it is automatically retained. Another way of improving number retention is breaking the digits into groups. The author recalled a nine-digit Social Security number after over a decade of non-use mainly from the way the digits are split into groups. First, a few numbers were recalled; then a single group is completed; eventually, all the numbers, and their distribution into the groups and sequencing came to mind. Visual images of the digits and their grouping helped.

It is noted in [31] that intelligence is genetic or hereditary since it reflects the quality of neural vibrations. Like any genetically determined physical characteristics such quality can be acquired initially through a mutant that eventually gets encoded genetically. Like any other physical characteristics such neural quality can be lost by atrophy. However, with rare exception, knowledge and intelligence come from training and experience and not genetically acquired.

The Nature of Thought and Knowledge

Thought is what happens in the mind and no one else knows what someone is thinking about. Therefore, we represent thought by some language and for purposes of science thought is represented by mathematical spaces, each one a language of some areas of science. This is the reason why most of the mathematics that has endured originated in natural science. Through sensation, synthesis of experience, training and study the mind creates physical concepts and laws of nature which, when suitably represented and systematized in the real world, becomes a physical theory that explains nature in terms of its laws. This understanding of nature that we now represent and express as physical theory is based on neural activities in the cortex expressed and stored as physical theory built on some laws of nature. Each field of science has a set of natural laws relevant to it. The new physics and its theoretical applications, for instance, are built on over 40 natural laws including biological laws and its application to the theory of intelligence, is built on 10 physical-psychological-biological laws [28, 29]. Present information says that the most developed cortex has barely 25% of its neurons activated or encoded. This means that human knowledge has vast potential for expansion. Moreover, even as the cortex is being fully utilized, evolution insures changes to provide greater capacity. That is the nature of evolution. Therefore, thought is *finite at any time but unbounded*. However, since an individual exists in finite space and time the reach of thought and capacity for knowledge is finite but unbounded. Moreover, the reach of our senses even after extension by advanced technology is finite. This fact has nothing to do with whether the visible universe is finite or infinite contrary to Olber's paradox that says infinite visible matter cannot exist since light would accumulate without limit at any point [74]. Furthermore, light has limited range.

Clearly, the mind can comprehend only finite set and measure because they are encoded by finite neural network. However, evidence shows that the mind has difficulty grasping what is large or small (depending on context). The introduction of scientific notation stretches that capability without bounds using order of magnitude but at the expense of accuracy since the digits are ignored in large or small number. Moreover, the mind functions dialectically. By negating what it can comprehend its domain of comprehension is further expanded but also at the expense of accuracy. The notion of infinity is the negation of finite and although the mind cannot comprehend infinity it can comprehend some aspects by negating some properties associated with being finite. We may define, for instance, an infinite set as one that cannot be contained in a finite set, *i.e.*, some element is left out each time we try to. Thus, the reference point for infinite is finite which the mind can grasp. For accuracy, we may define an infinite set as finite but unbounded since the mind can comprehend both "finite" and "unbounded", one being the negation of the other. However, there is a level of uncertainty or ambiguity in an infinite set since not all its elements can be identified, *i.e.*, its elements cannot be exhausted. Another example is well-defined nonterminating decimal. Being well-defined means that every digit is known or computable but not ALL of them are and that is uncertainty.

The mind is dialectical in another sense: its understanding of nature advances qualitatively, not through the sum of bits of information. For example, the mind is aware of empirical data. Through creativity and ability to recognize patterns, the mind may create and articulate some laws of nature based on the given data that would anchor a physical theory. A physical theory provides a lot more information than the sum of its component parts combined with empirical data. This is how the various disciplines of science are built.

Astonishing Natural Phenomena; Amazing Mental Capability

(1) Magnetic levitation is well known natural phenomenon. When bundles of strings of atoms of ferromagnetic material are suitably connected, aligned and oriented they produce a coherent vortex flux of superstrings called magnetic field. Then the material becomes a magnet. When a magnet is suitably light and shaped, e.g., thin disc with poles on opposite faces, it will lift up but wobble and fall unless it is stabilized by suitable weight or guided by the gyroscope. The lifting is due to flux-low-pressure complementary that creates low pressure around it in the Earth's gravitational field and pushes it outwards. This principle works on the flying saucer if it exists that can revolutionize travel in the Earth's gravitational flux that extends beyond the Moon.

(2) What about human levitation? This is also well-known although it is rare but the author knows a runner with the University of the Philippines' Varsity Team who claims he is partly aloft when he runs due to levitation. Recall that the atom is a magnet that has a vortex flux of superstrings producing a magnetic field that holds the orbital electrons together [43]. Since the human cells are made up of atoms they produce vortex fluxes within and around the body (called human aura, captured by Kerian photograph as bluish haze). Energy conservation requires that the latter be coherent, the atomic vortex fluxes eddies within it. In very rare cases, this flux is particularly intense and causes levitation, by flux-low-pressure complementarity. Being a physical characteristic, it is genetically encoded but can be acquired [31].

(3) Kerian photography is well developed in Russia. When a living thing, e.g., leaf or person is in a room Kerian photograph can capture the image after the leaf or person has been out of the room as if either of them was still there. This is only true of living things. Why? Living things have genes and they emit radiation that semi-agitate or even agitate dark matter that the Kerian photograph registers. Kerian photography also captures the human aura and this is now used in criminology. For example;

when a person dies normally, the human aura disappears in 24 hours; when there is trauma involved, e.g., struggle and foul play, the human aura may remain for as long as 72 hours. What is the explanation? Struggle and trauma disrupt coherence and impart kinetic energy to it that delays its normal conversion to dark energy by energy conservation.

(4) It is known that when a person experiences near-death situation he may acquire unusual skill such as sharp intuitions and extrasensory perception (ESP), telekinesis, unusual strength, facility with numbers and sudden skill, without training, at playing the piano or singing. Psychologists explain this in terms of the existence of "survival modules" activated during such traumatic experience. We interpret this as the existence of special neural network – a module –that gets activated by the experience. That module being a physical object is encoded in the gene.

(5) Telekinesis is a very rare but well-known phenomenon where a person bends a piece of metal by just being near it. The person who possesses this capability does not know why he has it nor is it understood why only very few individuals have it. We know that the CIR can focus, direct and propagate the energetic brain waves through the neural dendrite joins. The person who has this unusual feat simply has higher degree of this capability [30] encoded in his gene.

(6) No one has done a definitive scientific study on ESP among humans but there are scientific studies that show dogs possess ESP. The dog knows when its master is within 10 km on his way home. It may sit by the door and bark in anticipation of the master's arrival or go outside to meet him. The explanation is that the brain projects brain waves with specific vibration characteristic, i.e., electromagnetic waves encoded with appropriate neural vibration, through the dendrite tips just like radio waves sent out by the transmitter. By resonance, the brain waves vibrate the dendrite tips and, hence, the neural cells of the dog to cause sensation. Of course, such resonance is cultivated by proximity or touch. That is why it is not established with other dogs. Dogs being of the same species may have that built in resonance among them; a dog knows the presence of an approaching dog from kilometers away.

(7) Thought is not merely an abstraction but a material force or energy arising from physical processes in the brain. It can activate and run the motor of a prosthetic arm to move the latter as desired. It can also move the computer cursor, hit the icons and give commands.

(8) The mind is in total control of the body. With suitable preparation and conditioning one can walk barefoot on live ember or slow down an organ, e.g., the heart, and give it some rest. This is what happens in yoga. Others can lie back on a bed of sharp nails. Still others can hook a 100-kg load into his skin.

(9) Disuse of the body can cause atrophy. This applies to mental capability: it diminishes with disuse. In extreme cases, the person becomes senile through long period of mental inactivity; senility is a condition where the person acts likes a child.

(10) The opposite of disuse of body part is sustained, frequent and consistent use. This can induce genetic alteration and enhance that part of the body [31]. This is the effect, for instance, of constant training of an athlete.

(11) Proper training can turn an average mind into a superior one. In this author's experience there is a greater chance of training an average person to become a superior achiever. Moreover, sustained mental activity into old age not only avoids senility but also prolongs functional life. That is the stuff that makes an old statesman. Studies show that mathematicians remain productive in their 90s.

(12) There is ample evidence that stress diminishes effectiveness of the immune system, especially, with respect to cancer. Under normal condition, while the acquisition (mutation) and replication of cancer cells are kept at tolerable level by the immune system by destroying them at suitable rate, stress blunts this capability as the brain loses its grip on itself. This stress factor is not confined to cancer alone; it applies to other diseases, genetic and non-genetic, as well.

(13) Some inefficiency in the functioning of a body organ can be remedied by the mind. Proper training and suitable concentration, for instance, can provide immediate relief of constipation. Similarly, one can train himself to flap his ear or reduce the heartbeat rate to give the heart respite from lifetime work.

(14) There is what is known as Mozart effect. In controlled experiments one group of students listens to Mozart pieces on the eve of an examination and another group did not. The first group did significantly better compared to the other. This experiment has been replicated each time yielding the same result. Exposing the fetus to Mozart pieces also enhances the baby's mental alertness.

THE UNIFIED THEORY OF EVOLUTION

This section quotes [30, 31] liberally and is a theoretical application of GUT to biology that advances and updates Darwin's theory of evolution. Among the new elements in this theory are:

(1) Knowledge of how the gene produces the tissues of organism,

(2) Identification of the gene's medium for carrying out this function and

(3) Knowledge of how a foreign gene (mutant) replicates itself and becomes part of the pair of DNA strands in the chromosomes of the host organism. Item (3) explains the process of mutation. However, the most important contribution of this section is the discovery of several biological laws that explain not only the evolution of biological species but also the important functions of the gene in living things. The gene is identified as the crucial factor in the evolution of biological species but genetic alteration, the basis of evolution, is induced by the challenge to cope with and control the environment and gain dominance over competing species.

The Nature of Evolution: From Simple to Complex

There is overwhelming evidence that during the present ascendancy of our universe physical systems including biological systems evolve globally from simple to complex; local reverse cyclic processes and disintegration also occurs [30, 31]. This debunks the premise that the huge deposits of oil in some regions of the world came from decayed living organisms of the past. It is at odds with the nature of evolution since living organisms are more complex than carbon. Naturally, evolution of scientific knowledge follows the same track and debunks the idea that perfect knowledge existed in the past.

The uneven distribution of carbon follows from the law of uneven development [25, 42]. Carbon continues to form in the places where it has gained foothold, especially, in the Middle East, the Baku Fields of Russia, Texas, Indonesia and Venezuela, to mention a few. It also forms in the atmosphere. It was predicted 40 years ago that oil in the Middle East would be exhausted in 12 years. Not only was the prediction wrong, the huge deposit of oil in that region has continued to fuel wars to this day. While formation of any physical system is stochastic it is enhanced by the level of concentration of existing physical systems of the same kind (see relevant natural laws below). This is, of course, true of biological species described by the exponential growth function. However, this is also true of non-living physical system as an extension of energy conservation in the form of clustering as energy-conserving. This explains why crystals and pearls can be cultured.

Our universe evolved from the chaos of the Cosmic Burst [35, 39, 42] through its present complexity and higher order including biological order. Natural phenomena (visible physical systems) in our part of the Universe were nonexistent before the Big Bang, so were the natural laws they revealed. Now over 50 laws of nature have been discovered including biological laws some of which were introduced in [30, 31]. They explain natural phenomena and are verified by observation and experiments including data provided by the Hubble. They were synthesized from the wealth of observation and experimental data of the past and of the present rigorously formulated to anchor the development of GUT. GUT is articulated and stored in the language and logic of qualitative modeling [43, 78].

New conditions that arise open up new possibilities and directions of development of species. While the emergence of new species is stochastic in nature only stable physical systems survive and advance. This applies to formation of various atoms and molecules as well as biological species. For example, the high temperature and kinetic energy at our young universe just after the Cosmic Burst [37, 42] allowed only simple prima to form and only in cool cosmological bodies like Earth do complex atoms such as uranium exist. Even at this time, the powerful spin of and turbulence at the huge inner core of Jupiter keeps the temperature of that giant planet so high it is mainly gaseous. We summarize and articulate the content of this discussion as biological laws.

The Emergence of Living Organisms

Biological laws **BL1** and **BL2** below are essential for the emergence, evolution and extinction of biological species [28, 30, 31].

BL1 Stochastic Complexity Principle

Physical processes proceed stochastically in all possible configurations and towards greater complexity subject to the natural laws and already attained configurations and other boundary conditions.

While the emergence of biological species may seem spontaneous they are subject to natural laws, particularly, energy conservation and energy conservation equivalence that allows only stable configurations to survive. This is the reason man-made visible matter such as the prima (elementary particles) created at proton accelerators vanish in split second; they are unstable because nature does not need them, by the non-redundancy and non-extravagance principles [42]. At the same time, our universe takes tremendous amount of energy to alter and natural processes are difficult to tamper with even locally. For example, new species that have gained foothold, e.g., weeds, are difficult to eradicate for that would be intervening with energy conservation [42]. The next law is a companion law to the stochastic complexity principle above [30, 31].

BL2 Principle of Stability

Although all possible configurations of matter arise stochastically only the stable ones remain and are replicated; this is the main basis of evolution of biological systems.

In biology the capability attained by a species of living organism over hundreds of millions of years of evolution paves the way for further development. The brain is the principal factor for all processes in advanced organisms through brain wave generation and resonance [28, 29, 30, 31]. Nature has a number of equivalent expressions of energy conservation one of which is clustering since individual configurations of nature are chaotic, therefore, energy dissipating. Energy conservation induces matter to form cluster which may be initially random and chaotic, by the stochastic complexity principle, but stabilize towards higher form of order in complexity; this is the fundamental basis of evolution.

In an organism, energy conservation takes special forms: development of the immune system for protection against invading organisms, specialization of functions for greater efficiency, fractalization, reproduction, genetic encoding and evolution to higher order in complexity, to mention a few [28, 29, 30, 31]. Some micro-organisms develop resistance to medicines and plants develop toxins to ward off animals that feed on or harm them. This is how the poison ivy produced toxin. A species of the *cassava* root in the Philippines, substitute for rice as staple crop, has now the toxic chemical cyanide. The leaves of a species of *acacia* tree that giraffes feed on developed toxin and wiped out a whole herd of giraffe in the area [29, 31, 32]. Some plants developed sophisticated structures for survival and advancement, e.g., plants that have traps to catch insects. Another evolutionary advance was exhibited by the milk tree. The mother butterfly of a certain species lays its egg on the leaf of the milk tree but before doing so it goes around the leaf to make sure there is no other egg on it because the larva needs the leaf to feed on. In due course the milk tree suddenly put up a colorful module on each leaf that looked like the egg of the butterfly, obviously, to drive the mother butterfly away. Even more spectacular is the emergence of the carnivorous pitcher plant in the Philippines that has a pitcher-like trap that catches mice. The trap has enzymes that decompose the mice which is absorbed by the plant.

Some strains of malaria not only developed immunity to its cure –quinine – but even consume it. Some dreaded diseases thought to have vanished are back and thrive on anti-biotic that used to destroy them e.g., a strain of typhoid. New species of insects initially eradicated by chemicals like DDT have rendered the latter ineffective. This is the reason the use of anti-biotic drugs is carefully regulated to prevent surviving microbes from developing immunity to it.

The basic ingredient of plants and almost all living things is the carbon atom. Living organisms must have emerged when our universe cooled sufficiently to allow more complex clustering by carbon atoms. Carbon has four valence electrons and can bond with other carbon molecules to form huge molecules, e.g., protein, and the cells, tissues and organs of living organisms. This is the reason carbon-based compound is called organic. Carbon molecule has the strongest known bonding owing to its simplicity, symmetry, stability and the "rule of 8" principle of chemistry. In fact, when carbon is heated to high temperature under pressure and cooled quickly it becomes the hardest known material on Earth – diamond. Of course, there are rare known metal-based organisms found some 2 miles underground in the mines of South Africa that breathe pulverized metal for subsistence [5]; obviously, they cannot survive with even a mist of water but it is not known if carbon is one of its component molecules. Water is essential for almost all organisms. A considerable array of living organisms comes from magma that oozes out of cavities coming from the Earth's interior [107(a)]. This accounts for the fact that the volcanic islands of Galapagos off the coast of Ecuador have the most complete collection of biological species. This is where Darwin gathered his data for his theory of evolution in the years 1831 – 1836 [107(a)]. There are also living organisms in the ocean depths at boiling temperature around under-oceans volcanoes.

What distinguishes living from non-living things is the gene and the simplest known living thing is the virus consisting of protein strand with a gene on it. It is not an organism, of course, cannot reproduce and does not function like a living cell. Therefore, it needs a host organism to be able to reproduce by replicating its gene and becoming part of the DNA sequence in the chromosome of the host cell. When the genes produce the tissues in the cellular membrane the virus is reproduced as well.

The gene is a sequence of four nucleotide bases, adenine paired with thymine through hydrogen bonding and guanine paired with cytosine also joined by hydrogen. They form a sequence in the human body that makes up each DNA helical strand paired with another of the same composition, its mirror image. They form a parallel pair coiled into a helix in the chromosomes inside the nucleus. The genes of a living thing such as plant, virus or organism, determine its characteristics. This is well accepted in genetics and we state it as our first biological law (applies to biological species alone).

BL3 The Determinant Law

The gene determines every physical characteristic of a plant or organism; conversely, every physical characteristic of a plant or animal is determined by a gene.

What is not known is that the genetic content of a body can be modified or altered.

The Internal Basis of Evolution

We first describe biological phenomena involving the cell and formulate the natural laws that explain them. A cell is altered genetically (mutant) when its gene is modified or a new gene is added to its DNA helical strand. The new or modified gene immediately forms its mirror image in the other DNA strand of the pair and replicates itself quickly, locally at first, altering the genetic content of cells in its immediate neighborhood and producing the corresponding characteristic there. Then the genetic make-up of the organism is altered over time unless the mutant is destroyed by the immune system. Genetic alteration may occur through direct exposure to a foreign gene or indirectly through mental pre-occupation with an object (living or not). Exposure to radiation or carcinogenic chemical may also induce mutation. The man whose face was covered by the black hair must have been exposed to the wolf's gene directly when he was still a fetus. Birthmark usually appears through mental preoccupation by the mother with some object during conception even without exposure to a foreign gene. There are ample documented cases of birthmark resembling an object, living or not, due to the pregnant mother's preoccupation with or craving for it during the first few weeks of pregnancy. There was a case of a pregnant mother who craved for cooked catfish during early pregnancy. The baby girl came with a pair of bluish arcs issuing from the corner of its mouth that looked like the catfish's pair of whiskers. There was no exposure to a living gene in this case.

In the rural Philippines dried rice grains were pounded by a heavy pestle (pole) on mortar to separate seeds from the husk. This leaves permanent thick discolored markings on the palm which means that the markings had been genetically encoded so that in the normal turnover of cells the new ones that replaced the old cells carried that gene. Fortunately the last two cases only involve slight discoloration which may disappear in a couple of decade to atrophy.

Genetic alteration may also occur through consistent, sustained, frequent and repetitive use of body part. This is the scientific basis of drills (e.g., muscle building and athletic drills) to develop the muscles of certain parts of the body. In other words, a mutant may be encoded genetically over time. A similar situation must have occurred in the evolution of species that walked on four legs and became erect. Occasionally, members stood up to see farther in search of food or spot an intruder. Over time, this new physical characteristic of standing erect becomes encoded genetically until standing became more permanent. Then the front legs find new functions and the anatomy of both the front legs and the face in response to the new situation genetically encoded. The prairie dog that can stand erect for sustained has reached this phase in its evolutionary advance.

Rice farmers on plain fields in rural Philippines till the field by walking on mud, sometimes waist-deep. In time, some farmers developed flat foot, nature's way of coping with the environment since being flat-footed prevents sinking deep into the mud. The genetic effect is to thicken the appropriate tissues of the flat of the foot which is encoded in the gene. In due course, younger children of flat-footed farmers usually inherit this condition.

In hilly fruit farms farmers climb slippery hill or coconut tree by sticking their big toes into the soil or cuts on coconut trunk. Some of them develop oversized, protruding strong big toes that are slightly arched away from the little toes, a condition usually inherited by the younger children. The last two examples of genetic alteration do not vanish through atrophy.

Technically, a mutagen is an agent of mutation such as radiation or some chemical. Since a gene can alter the genetic make-up of other cells we also call it a mutagen if it does. Similarly we call a cancer cell or its gene a carcinogen. We capture the above phenomena by the following biological law.

BL4 Law of Genetic Alteration

The possible sources of genetic alteration are: radiation, e.g., cosmic waves, direct exposure to some genes, frequent, consistent and sustained use of body part, irritation and pre-occupation with or craving for some object.

Although the embryo is most vulnerable to genetic alteration, young adults also experience it. For example, an adopted child who has no blood relationship with his adoptive parents may later exhibit physical features of the latter. Young couple develops similar features later in life. One implication is: new physical features (encoded in the gene) including pre-disposition to some diseases may appear at any point in the family tree. Once a person has acquired a genetic characteristic the gene responsible for it cannot be removed since it is present in every cell. However, it can be modified or altered to neutralize undesirable physical characteristics it gives rise to. Moreover, disuse of the affected body part may lead to atrophy or reduced prominence of its physical characteristics.

Genetic modification has medical application. For example, cystic fibrosis can now be treated by inserting a neutralizer gene into the cell. This is done by infecting a virus (e.g., common cold virus) with a neutralizer gene and having it inhaled by the patient. Then the virus enters the cell of the patient and replicates its genes there including the one just acquired. This technology is now possible for the treatment of other genetic diseases since the neutralizer gene will replicate themselves in every cell of the patient in due course regardless of where it is implanted. Recently, there has been an advance in this technology; DNA can now be injected into the body and cultured without need for a virus lever.

When a person is in pain he may cope with it by modifying posture, e.g., when the pain is in the spinal column, he adjusts vertebral arch to ease the pain; when this is sustained for suitable period of time corresponding genetic alteration that sustains and encodes this new physical characteristic (arched vertebra) occurs. The distortion is not only permanent but also genetic or hereditary.

Given all this information we raise these important questions: (1) How does a foreign gene replicate itself in the chromosomes of the host cell that becomes part of the latter's DNA? (2) How does the altered gene spread to other cells? (3) How does the gene produce the tissues of a living organism?

It is popularly believed that the altered gene is carried by the mutant cell through body fluids. Assuming this is true, how does it replicate itself and become part of the DNA sequence of other cells and contribute to the genetic make-up of the individual?

We look at other body processes and find the common thread that runs through them. We recall that that every piece of matter vibrates due to the impact of cosmic waves and its normal vibration characteristics are determined principally by its internal structure [30, 31]. Their nuclei generate basic cosmic waves with these characteristics. In a sense organ visible signals from an external event add new component characteristics to the nuclear normal vibration that get encoded on the brain waves they generate. For example, the taste bud is the receptor of taste, its base the bundle of nerve endings connected to the cortex by linear sequences of nerve cells. By resonance, molecules of food vibrate the appropriate taste buds and the nerve endings at the base on the sides or tips of the tongue, adding new components to the basic cosmic waves generated by their nuclei and turning them to brain waves (encoded with information or vibration characteristics). The information is transmitted to the cortex through the nerves for interpretation and encoding as memory. At this time, the individual experiences the sensation of taste in the creative-integrative region (CIR) of the cortex [30]. In this case, the taste bud is both receptor and converter. This information is the basis of making synthetic foods and sugar-free sweets for diabetics that retain their tastes using carbon as base molecule without the undesirable side effects. For example, sugar-free sweets for diabetics that are much sweeter than real sugar do not have the viscosity of sugar that damages the fine blood vessels of the kidney. The same principle applies the manufacture of various synthetic foods such as synthetic beef and crab meat. Thus, synthetic foods can be designed and manufactured to suit the palate.

In hearing the eardrum is the receptor. Immersed in cochlea fluid are three tiniest bones of the body – stapes (stirrup), incus (anvil) and malleus (hammer) – arranged in a triangle for detecting the direction the sound is coming from (much like the way authorities pinpoint the location of illegal underground radio station). Lining the wall of the cochleal canal are strands of hair with different vibration characteristics. Each strand resonates with encoded brain waves of similar wave characteristics. Audio signals from the outside vibrate the eardrum which vibrates the malleus connected to it with the encoded signals. The malleus hits and vibrates the incus that, in turn, vibrates the stapes that generates and propagates encoded brain waves. They resonate with the corresponding hair strands with similar vibration characteristics which then resonate with the nerve endings at their follicles that transmit the signals to the CIR for eventual encoding in the appropriate sensation regions as components of memory. Thus, in hearing the eardrum is the receptor and the hair strands are the convertor (for details on the functions of other sense organs, see [28, 29, 30, 31]).

Encoding in the CIR involves activation of neural network interconnection that vibrates with corresponding characteristics. When brain waves stop coming, the network de-activates and reverts back to normal vibration. When suitable cosmic waves hit the network (coming from the CIR or directly from the Cosmos) it is activated and vibrated with exactly the same characteristics as the ones encoded in it. Then the encoded information is recalled. Thought is due to vibration of appropriate neural network encoded with information. The network propagates brain waves (encoded with thought) through the dendrite tips (which act like radio or TV transmission antennas). During sleep components of encoded neural network are transmitted and dispersed, by resonance, to appropriate sensation regions. When needed for thought they are recalled and re-composed in the CIR [29]. Such thought processes are subject to energy conservation [29].

Clearly, brain waves (basic cosmic waves wiggled by atomic or molecular vibration) are the medium of communication between the brain and every part of the human body, especially, the sense organs. It is also the medium of the brain to command other parts of the body. Conscious commands come from thought by the CIR (e.g., an athlete executing a game) but some commands are executed by the automatic system, e.g., reflex. The thought of lemon, for instance, triggers salivary secretion automatically. Our eyes blink automatically when an object is thrown at us so do we pull our finger out of ember.

We now consider the gene's medium for its function. This much we know: (a) the gene produces the tissues of plants and animals and (b) vibrated by energetic brain waves that cause pain sensation, the neural cell produces molecule in the cellular membrane [76, 84]. The latter was verified by experiments at the Pavlov Institute of Physiology [76, 84]. We make the following conclusions:

(1)   By the principle of non-redundancy and non-extravagance brain waves are the only physical systems that agitate and convert superstrings to prima that form the molecules observed in [76, 84].

(2)   The gene produces the molecules that form the tissues and chemicals of a living organism. From (1) and the principles of non-redundancy and non-extravagance, it follows that the gene also produces

(*i.e.*, converts from superstrings) the tissues and chemicals by its brain waves in the cellular membrane.

(3) By the principle of non-redundancy and non-extravagance, it follows from (2) that it is the gene that produces the molecule in the neural membrane observed in [76, 84].

(4) Although verification that brain waves produce molecules in the cellular membrane has been done only for brain waves encoded with the vibration characteristic of pain sensation, by the principles non-redundancy and non-extravagance, brain waves produce all the molecules that form the tissues and chemicals of the body in the cellular membranes.

To summarize, the gene produces the chemicals and tissues of plants and animals in the cellular membrane using brain waves. Similarly, thought commands through suitable genetic vibration that generates brain waves, any part of the body to do its functions. We, generalize this to other functions that include the gene's capability to generate and propagate brain waves in all directions that reproduces its mirror image in the x or y chromosome of each cell and produces the tissues and chemicals in the cellular membrane in the right place. By the principles of non-redundancy and non-extravagance it is the same capability that allows a new gene or mutant to reproduce its mirror image in the chromosome of the host cell. This occurs when there is discordant resonance with a gene of the host cell. It results in replication of the mutant gene and alteration of the genetic make up of the host cell. When there is resonance with the vibration of some gene in the host cell the wave or vibration characteristics of one superpose on the other resulting in mutual genetic modification that produces a composite gene with its composite brain wave characteristics coming from the interacting brain waves. We generalize the law of genetic alteration with more components that have medical applications for treatment of genetic diseases.

BL5 Law of Induced Mutation, Genetic Alteration and Modification

When a new or foreign gene is not at resonance with a gene in the host cell its brain waves agitate the chromosomes of the host cell and produces its mirror image there in one DNA strand that replicates itself in the opposite strand to form a mature gene (i.e., a pair). When the brain waves of the foreign gene resonates with a gene of the host cell the characteristics of one superpose on the other and produces a composite (modified) gene that projects its composite brain wave characteristics.

Alteration of a gene by a foreign gene is localized at first to the neighborhood of the altered gene where the physical characteristic it determines appears and develops until it spreads to other parts of the body. We note that this does not apply to normal genes of the body since the cells have the same genetic make up and there is full resonance among the brain waves they generate. Therefore, this kind of genetic agitation or modification does not occur. (For details about brain wave generation see [30, 31])

Apparently, the genes of certain species, e.g., primates, generate intense brain waves that induce birthmark on the human embryo. This accounts for the prevalence of birthmarks with physical features of primates. Brain wave propagated by a gene (in all directions) has just enough energy to produce a gene in one of the DNA pair of strands it hits in the chromosome of a cell (a form of quantization principle). Then the latter gene produces almost instantaneously its mirror image in the other strand. We summarize our discussion as a biological law:

BL6 Genetic Brain Wave Propagation and Encoding

In an organism the gene generates and propagates brain waves in all directions; it is in harmony and at resonance with all its genes. A foreign gene or mutagen may have discordant resonance with a chromosome, insert its mirror image into one of the pair of DNA strands of the recipient cell and produce its mirror image on the other DNA strand of the pair to produce a mutagen.

Encoding a mutant is completed when every cell has been similarly altered genetically. Then this characteristic becomes hereditary. First, a gene is altered (becoming a mutagen) then other genes are altered as well, locally at first, until this alteration spreads to the entire body. One implication is that some genetic disorder may be acquired in a later generation. Another is that once a new gene has been encoded it cannot be removed. Its effect can only be neutralized. (However, the physical feature it determines may disappear by atrophy). This is the principle used in the treatment of cystic fibrosis using cold virus as agent. Since a virus must enter a cell to reproduce itself, it immediately induces genetic alteration in the host cell by the neutralizer gene which quickly spreads to the lungs and provides remedy. As we shall see in the last section this technique has universal application for treating genetic defect including cancer although the more appropriate treatment for cancer is genetic sterilization.

Corollary

A foreign gene from the same species may, by resonance, superpose its brain waves on the brain waves of some gene to form composite brain waves that modify that gene accordingly.

We assume that the younger the organism the more vulnerable it is to genetic alteration, the embryo being the most vulnerable. There is plenty of evidence of gene alteration when there is close proximity between a young person and an adult or even among young persons. An adopted child generally takes some features from the adoptive parents. This observation is at the phenomenal level. Geneticists should try to verify this experimentally, *i.e.*, make case studies of adopted babies and find out if there is any genetic alteration (appearance of genes from adoptive parents) when they become adult. The cosmetic implication of this phenomenon is quite obvious.

Since radiation, like brain waves, is quite energetic, it can also induce genetic alteration directly. Then the cell with this altered gene becomes a mutagen, generates brain waves of specific characteristics and causes genetic alteration in normal cells. This is the basic mechanics of cancer. It has been noted recently [1] that the onset of cancer follows the standard dynamics [35] which has three phases: (1) the phase of evolution from order to chaos (normal cell), (1) bifurcation or multifurcation of normal process, that is, the transitional phase of chaos where new characteristics appear, and (3) the phase of evolution from chaos to order (cancer in progress and replicating its gene in the normal cells). Phase (2) is genetic alteration phase. One of the physical characteristics of cancer is rapid reproduction into lumps that create pressure and damage the tissues or organs.

The explanation of gene alteration including some birthmarks through direct exposure to the genes of adoptive parents or other species and contact with some chemicals is now quite obvious. We deal with the less obvious ones such as development of flat-foot, strong, oversized and protruding big toes, developed strong muscles of the weight lifter and reflexes and genetic alteration without direct exposure to brain waves from foreign genes but results in mutation such as the above example of bluish replica of a catfish's whiskers issuing from the corners of the baby's mouth.

The first category of gene alteration such as flat-foot, etc., involves frequent, consistent, sustained and repetitive use of some part of the body where the genetic alteration is to be introduced. It agitates its chromosome and induces conversion of dark matter to the base components of the gene to alter the cell genetically and produce the corresponding mutant. The same genetic alteration occurs in the case of a reflex except that the altered cell that results from it belongs to the automatic system and the new gene determines some physical skill such as driving or playing the piano. This has implication for teaching these skills. It is important that training be perfect by correcting every error especially at the initial phase to satisfy the requirement of frequent, consistent and sustained motion to induce the required genetic alteration. This is also true of such precise disciplines as mathematics and science, especially, in the teaching of mathematical and scientific reasoning even if no genetic alteration is involved since they are not physical characteristics. Instead, clear and consistent neural network interconnection is involved [29, 31]. Thus, what is learned becomes mental reflex. For example, an expert typist feels it when he hits the wrong key so does an English teacher when there is error in grammar or idiom (details in [29, 31]).

In the case of the arched spinal column the genes in the chromosomes of affected cells propagate brain waves whose characteristics capture its structure and induce, via discordant resonance, the corresponding alteration in DNA gene sequencing and characteristic vibration. The geometrical configuration of the physical characteristics by a gene translates into the appropriate sequencing of base molecules that make up the gene. The genetic activation law governs actualization of its physical characteristics at the same place.

Genes are replicated in the normal turnover of cells; this is what happens in healing wounds, normal oxidation and ageing. Therefore, every physical characteristic, old or mutant, is retained (unless the area where it shows is removed early enough). Moreover, a mutant need not wait for the next generation to show or move physically to spread; they spread by sending out discordant brain waves that replicate their genes in the chromosomes of the other cells of the body. However, body fluids enhance the spread of cancer by distributing the cancer cells in the body and facilitating genetic alteration. Furthermore, a gene may be acquired by exposure to gene-generated brain waves from other organisms.

When a person loses an arm will it induce genetic alteration? Yes and no. No, because there is no source of discordant resonance to cause agitation of the chromosomes that can produce a new gene. Removing a gene does not involve discordant resonance or agitation; therefore, it does not induce genetic alteration. Yes, because the other arm takes over the functions of the lost arm and works overtime. Then this arm takes new characteristics such as bigger and stronger muscles and developed skills in accordance with the induced mutation, genetic alteration and modificaton law.

BL7 Direct and Indirect Genetic Alteration Law

In an organism genetic alteration may occur through direct exposure to radiation including cosmic waves of suitable intensity and gene's brain waves, direct contact with some chemicals, sustained use of body part involving repetitive motion and sustained concentrated thought about a material object.

In mental pre-occupation or craving for something such as cooked catfish to eat during the mother's early pregnancy concentrated thought generates intense energetic brain waves having the characteristic vibrations of the object of thought that can agitate and convert dark matter to genetic materials that may cause genetic alteration. When they hit and agitate the chromosomes of the mother's embryo at the onset of pregnancy they produce the genes with those vibration characteristics and the mutants are genetically encoded there.

Survival and Advancement of Biological Species

Three biological laws govern preservation and advancement of species: (1) passing the genes to the offspring, (2) insuring body parts form in the right places and (3) in the competition for survival the species is able to modify itself to be able not only to cope with the challenges from the environment but also to gain more and more advantages over and dominate other species. With respect to (2) it is governed by the genetic activation law below.

Once activated the genes convert superstrings to molecules that form the tissues of the specific part of the body together with its other physiological and genetic components in the cell membrane [76, 84]. Malfunction in this department is responsible for growth of, say, an extra finger on one's hand that involves genetic alteration or mutation.

A gene turns on only in the right place to convert superstrings to suitable molecules on cue from the environment. This is how the cells, tissues and other chemicals are produced in the various parts of the body which are determined by the genes responsible for their formation and development. The specific environment is determined by enzymes and distribution of cumulus cells that initially surround the egg cell. Here lies the mechanism for preventing an organ from growing in the wrong place, say, an eye growing on the tip of the nose. Moreover, even the turn over of cells, *i.e.*, replacement of dead cells by new cells, are genetic in the sense that each replacement has the genetic code of the oxidized old cell. We synthesize this observation as another biological law.

BL8 Law of Genetic Activation

Activation or non-activation of a gene is controlled on cue from immediate environment; this prevents anomaly in the development of an organism.

The ability to reproduce itself and pass on its characteristics to the offspring is the distinguishing feature of a living organism. The virus does not have this capability. Therefore, it enters the cell of an organism and alters the latter's genes to reproduce itself as a new characteristic. In effect, once a virus enters the body it automatically becomes a mutant embedded in the host's cell. It gets replicated in the tissues that the normal gene reproduces. However, since it does not reproduce itself it is eventually destroyed by the immune system, if the latter survives, without producing an offspring.

From the simplest to the most complex living thing beyond virus, reproduction is, essentially, cell division. The only new element is fertilization in plants and animals but even this is quickly followed by cell division that results in the embryo of a plant or animal.

The agent of reproduction is the gene, a subunit of the DNA that form the chromosome (x and y in male, x and x in female), in which is encoded as chain of four nucleotide bases a particular characteristic of the organism. The sequencing is part of the program for reproduction for it determines the configuration of the molecules and enzymes that form the tissues and organs of the organism. The totality of the genes in the DNA of an organism, arranged in sequence of base materials that form the genes, encodes the entire structure and physical characteristics and behavior of the organism. In a human female each of the two DNA strands in the pair is in the x-chromosomes but for male one is in the x-chromosomes and the other in the y-chromosomes. (It is conjectured that being right- or left-handed has something to do with the relative position between the x- and y-chromosomes in the male and in the female between its x-chromosomes and the x-chromosomes that comes from the male)

A matured DNA is a pair of such genetic sequences, one being the mirror image of the other within the double helical strand. The helical DNA configuration in a complex organism is an optimal balance between space and efficiency of replication. The only topologically feasible DNA symmetric replication without crossing a coil during cell division is for the double helix to split longitudinally along the helix between the two pairs of DNA strands, half of the pair slightly contracting as inner coil and sliding out of the outer coil into the offspring cell. Then in each offspring cell the DNA replicates its mirror image immediately through the brain waves its genes project to produce its parallel DNA strands in the chromosomes and form, again, a matured DNA pair. Male or female, the mirror image of the genes in the daughter cell is in the x-chromosomes since if the x-chromosomes from the male goes to the daughter cell its DNA there will form its mirror image in the x-chromosomes of the female (to produce a female offspring with x-, x- chromosomes); so does the DNA in the y-chromosomes that goes to the female (to produce a male offspring with x-, y-chromosomes).

The egg has half the x-chromosomes of the female containing one of the pair of the DNA strands. Each of the male's sperm cells has either the x-chromosome with its DNA strand or the y-chromosome with the other DNA strand of the pair. Therefore, it has half the chromosomes of the matured cell.

Fertilization occurs when the sperm cell penetrates the egg cell and inserts its x- or y-chromosomes containing a full strand of DNA in the egg's x-chromosomes to pair with the latter's helical DNA strand and form a full parallel pair of helical DNA strands. The process of pairing is the reverse of mitosis, the helical half-chromosomes of the sperm inserting itself into the helical half-chromosomes of the egg, expands slightly to fit the female helical DNA so that the end result is a matured cell containing the pair of parallel helical DNA strands in the x-, x- or x-, y- chromosomes. We assume this process quite precise so that, by resonance, a gene from each partner that will produce the tissues of the same part of the body of the offspring will form a pair, one on each DNA strand of the helical pair of the fertilized egg. Their brain waves superpose on each other to form composite wave characteristics and modify each other accordingly so that they project the same composite wave characteristics. This modified gene pair determines the characteristics of that body part of the offspring. If the brain-wave-characteristics of a partner's gene are more prominent than those of its pair then this partner's physical characteristics will be more prominent in this offspring's body part. Genes of one partner that do not have counterpart in the other produce their respective mirror images on the DNA strand contributed by the other partner so that they are shared by the offspring. One partner's gene is dominant over the other partner's gene if its projected brain waves are slightly more intense so their composite brain waves have more dominant characteristics coming from the former. The component physical characteristics determined by a gene are present in each partner only that one is more prominent in one than in the other.

Each of the male and female genitals has counterpart genes on the other but one set is prominent over the other. For example the male genitals have counterpart components in the female genitals the difference being that the male characteristics are dominant in the former (and vice versa). In rare cases (e.g., hermaphrodite) they are equally prominent so that the individual has both penis and vagina. Dominance is sustained and expressed in feelings, emotions and personality make-up by the appropriate hormones (also genetically determined) but hormonal imbalance can erode them. Moreover, the levels of hormones and enzymes and the nature of the physiological process involved in the development of the fetus are genetically determined. It is conjectured that being gay is due to lack of clarity and prominence of male gender characteristics and they are also genetically determined. In other words, being gay is genetic. However, since genetic alteration is continually induced by the environment "gayness" may appear at any point in the family tree (similar analysis for lesbian).

Fertilization quickly triggers formation of gene mirror images and composite genetic brain waves and corresponding genetic modification. Since it involves release of energy as brain waves it de-stabilizes the fertilized egg that is resolved by mitosis. Mitosis conforms to energy conservation in terms of symmetry of end result, namely, the replication or fractalization of cells with shared structure.

BL9 Law of Reproduction and Correspondence

During fertilization the single full DNA strands from the sperm and egg are paired, by resonance, so that (1) each pair of genes that produces the same part of the embryo are opposite each other in the chromosomes of the fertilized egg; (2) their brain waves superpose on each other resulting in common composite brain waves that modify the pair by resonance; (3) unpaired genes from each parent DNA produce their mirror images in the other to complete genetic composition of the fertilized egg and (4) the energy involved in (1) – (3) causes instability in the fertilized egg that triggers mitosis and starts the development of the embryonic offspring.

The modification of the genes during fertilization is preserved by mitosis and further development of the embryo. Clearly, normal reproduction is already a factor for genetic alteration and evolution.

In humans the embryonic cells are physically identical during the first few days of cell division. Differentiation begins with changes in nuclear instruction in the use of the genes by turning them on, *i.e.*, translating genetic codes to produce appropriate tissues and chemicals or keeping them turned off temporarily for some and permanently for others which are excreted when the cells die. The changes are triggered by environmental signals around the cells, e.g., presence of enzymes, hormones and cumulus cells. For example, the genes that will produce the tissues of the embryonic heart will turn on and produce the right molecules in the membranes of the cells there including the base molecules of genetic materials where the genes will be replicated along with the cell division that will build the embryonic heart. All the cells there will cell divide genetically and replace the dead cells. The genes in the embryonic heart with instruction to produce the pancreatic tissues are permanently shut off. The whole process is restarted in the chromosomes of the matured offspring.

During the period of growth of an individual the cells multiply (by cell division) exponentially and die linearly (mainly through oxidation). When the individual reaches full growth the growth function tapers off to a linear function. During the ageing process the linear growth function takes a lesser slope than the cell death function and the individual shrinks and functioning neural cells decline. This is enhanced by sustained direct exposure to sunlight. However, there is therapy for countering the ageing process and avoiding senility, at least, mentally. One is to keep the mind active. Anti-oxidants minimize cell oxidation so does regular bath with cold water.

Groups of bases adjacent to the coding sequences affect the quantities and disposition of gene products; in higher organisms these non-coding sequences outnumber the coding sequences 10 to 1 or more. Their functions are essentially unknown at this time. However, this author takes the position that they serve as landmarks distributed suitably and turn on the genes to produce the tissues of organs and other parts of the body in the right places in accordance with the genetic activation law (see [30, 31] on physiological processes involved in mitosis).

Reproduction and evolution are intimately related. The present reproductive precision took hundreds of millions of years to achieve by the combination of stochastic complexity and energy conservation laws. Trillions of different structures emerge from this fractal process, ineffective and unstable ones discarded outright and replaced by stable ones, until present structures and level of precision are achieved which, even now, are not perfect yet. Numerous genetic defects like mental disorder, cystic fibrosis, diabetes and predisposition to certain diseases still confront the human species. Moreover, stress during the development of the fetus may result in anomaly. At any rate, evolution is a great experiment and only the most stable species able to cope with the challenge posed by the environment survive.

The reproductive process has practical applications in agriculture, especially, plant breeding. In medicine scientists are able to isolate a suitable gene; then a virus infected with it is injected into the patient to alter the genetic code and neutralize defective gene, the beneficial effect raised a thousand times, thanks to the Genome Project, it is now possible to remedy genetic predisposition to any disease. The genome project has mapped the 25, 000 – 30, 000 genes of the human body. In a few years, they can be examined for defects that can be neutralized. Now molecular biology and genetics can engineer some parts of the human body called stem cells, e.g., arm or finger, by applying the genetic activation law.

Cases of flat foot, strong oversized big toes and developed muscles of the weight lifter show advancement of species through genetic alteration induced by the environment. On the other side of the coin is disuse of body parts that leads to atrophy over generations of the species. We cite a dramatic example of this development: evolution of the snake.

Some species of the animal kingdom have done spectacular evolutionary experiment. All species of snake experimented with a pair of legs millions of years ago, located at a point 1/3 of its length from tip of tail. We may then assume that it had a stiff body but with such elongated body it must have been clumsy to move around with the two legs. Like the farmer who developed flat feet and oversized strong big toes and the weight lifter with strong developed muscle, in accordance with BL4, the snake must have moved to attain some body flexibility and, over a long period of time, acquired agility that enabled it to wiggle through bushes and climb trees with ease. At that time the legs had already atrophied completely, discarded and replaced by wiggling speed and agility, that only their scars remain (still visible in any snake today). There was a nodal point here: splitting of species. Some discarded speed in favor of greater size, weight and muscular strength needed not only for defense but also for breaking the bones of their prey (the python and anaconda still have them). However, huge size and weight became a disadvantage: huge amount of food is needed to sustain the species. Over the next several million years the big ones joined their agile and speedy cousins and discarded muscles in favor of the venom (apparently, only the pythons and anaconda remain but they have no competition now). Smallness allowed them to hide from predators in holes and crevices. Security was further augmented by the venom both as a weapon against predators and for digestion. Almost all species of snakes joined this last bandwagon.

The snake has eyes but it is blind. Apparently, the tongue that is sensitive to vibration is a more effective sensor.

Plants also went through the same evolutionary advance. We can look at poison ivy as having developed toxin to ward off organisms that harm them. Then there was the case of acacia trees that developed toxins and killed and wiped out a herd of giraffe that pastured on their leaves (reported by National Geographic).

A dramatic evolutionary move was displayed by the milk tree. The mother butterfly of a certain species lays its egg on the leaf of the milk tree. However, before, laying its egg it goes around the leaf to make sure no other egg is there for its larvae spectacular to consume the leaf. Then, suddenly, the milk tree produced colorful modules on each leaf that looked like the butterfly's egg, obviously, to drive the mother butterfly away. The only possible explanation based on our theory is that the genes of the butterfly induced suitable genetic alteration that produced the physical characteristics of the modules. (Facts reported by National Geographic).

We capture this account as a biological law.

BL10 Law of Advancement and Atrophy

The need for survival and advancement and overcoming the challenges from the environment and competition from other species induces suitable genetic alteration that allows species to prevail.

Some aquatic and land organisms developed varied ways of catching prey; some plants developed suitable structures to trap insects and tiny animals like mice, e.g., the pitcher plant in Palawan.

We can see here that the species response to the environment is not merely a defensive action such as coping for survival but an offensive action as well: to gain dominance over other species.

We referred to instincts as guide to evolutionary development. Like reflex they are encoded in the genes. For instance, the instinct to hunt is encoded in the lion's gene but in the division of labor in the clan it is the mother lion's task to hunt and train the female cub to actualize this instinct. It borrows the prey's cub, e.g., a deer's, for training and returns it at the end of the day to grow, mature and become a prey. The male lion guards the territorial boundary (about 650 square km) and confronts intruders. An outsider may pass through provided it submits to the lion's authority by lying on its back. Lions and most species of the animal kingdom have very strict code of ethics and strong social organization.

The instinct for survival of species is the principal factor for the development of appropriate anatomy and functions for preservation and advancement of species. For instance, it has been reported by National Geographic recently that in an all-female cow herd one bore a cub through self-fertilization. We capture our discussion in another biological law.

BL11 The Law of Evolution and Mutation

Diverse action of cosmic energy and the environment, particularly, the need for survival and advancement, induces infinitesimal changes in configuration of organism which, together with combination of genes from parents, lead to infinitesimal evolutionary changes in species that may not be observable but add up to significant alteration in the genes and corresponding physical characteristics over a long period of time.

Finally, we introduce a natural law that is particularly important for the treatment of genetic diseases.

BL12 Law of Resonance and Superposition

Two waves of the same order of magnitude resonate have maximum resonance when they have exactly the same characteristics and the principal determinant of level of resonance is wavelength or frequency; when two waves resonate their wave characteristics superpose on each other and produce a composite wave.

The law of resonance was discovered in the analysis of the disastrous final flight of the Columbia Space Shuttle in 2003 [40].

These biological laws along with the general laws of nature that we have identified form the axioms of the theory of evolution as a sub theory of GUT.

Outstanding Issues

There are issues regarding the origin and nature of biological life, particularly, human life. Huge resources are being spent in attempts to resolve them. For instance, there is the Mars probe to find out if there is life in that planet and if there is, to shed light on how life started here on Earth. Traditional anthropologists believe that human life started in Africa and life elsewhere is the result of migration. Others believe that the seeds of life were implanted here on Earth from outer space, perhaps, carried by asteroids or meteors or tails of comets. There is yet another view taken by the Raelians, a cult based in South Korea, who believe that human beings were bio-engineered by aliens. To them we pose this question: who bio-engineered the aliens?

Given that our universe is made of the same stuff, the superstrings, and their conversion to visible matter, the prima, why should life emerge only here on Earth and only in some region of the Earth? Why should human life emerge only in Africa? If other biological organisms originate in many places on Earth, why should not human beings, too? Why should intelligent beings in this vast universe of ours exist only in our planet? Why should biological species be based only on carbon (some species are metal based [10])? Why not some other element elsewhere in our universe? We leave these questions for us to ponder.

THEORY OF CHAOS AND TURBULENCE

We shall develop a physical theory of turbulence. We have already defined chaos and turbulence and we look at chaos as transition from one type of turbulence to another in what we call the standard dynamics [35] that has three phases: phase 1, the phase of normalcy where a physical system is in normal motion, e.g., normal vibration; phase three, the transitional phase of chaos; and phase 3, the phase of turbulence. For example, we can take phase 1 as a calm summer day in the Pacific. Suitable under-ocean volcanic activity creates suitably broad contiguous warm ocean surface that causes low pressure or depression over it. It sucks the gas molecules around it in their trillions and their collision creates the second phase of chaos or Phase 2 since none of the colliding molecules can be monitored or its path predicted and yet every molecule is subject to natural laws. Thus, we have a typical case of mixture of order none of which is identifiable. However, this phase is energy dissipating due to collisions so that by energy conservation the sucked gas molecules evolve an order called vortex, in this case, a cyclone like hurricane. Therefore, this is a coherence of order called turbulence or phase 3 of the standard dynamics. Among the familiar examples of the standard dynamics are weather turbulence and the stock market. Quantitative modeling alone has not saved the weatherman from frequent embarrassment and the stock broker from occasional ruin. Thus, this type of problem of turbulence is unsolved. The irony is: the solution might keep the weatherman and stockbroker jobless.

Chaos, turbulence and fractal are generally lumped up together in the scientific literature; unfortunately, they are not equivalent dynamics but distinct phases of the standard dynamics:

(1) Evolution from order to chaos,

(2) Transitional phase of chaos and

(3) Evolution from chaos to order called turbulence.

Then turbulence may or may not fade away depending on whether the condition for its existence is sustained. For example, Jupiter has a cyclone that has been there for 300 years. The usual examples of standard dynamics are weather and financial market dynamics [53, 109]. The standard turbulence problem is: find a mathematical model of the standard dynamics that has predictive capability on its future state. While there is great effort at trying to find the mathematics of the turbulence problem in the stock market a solution will also wipe it out for good as market fluctuation grinds to a halt. The common thread of this dynamics is uncertainty – local and infinitesimal at all phases and also global at phase (2) – which places them under the category of ambiguous physical system (involving large set and numbers). Rapid infinitesimal local motion induces uncertainty at phases (1) and (3) (e.g., Brownian motion of gas molecules) similar to uncertainty at limit set of fractal.

We focus on earthly turbulence: earthquake, volcanic activity, typhoon, tornado and lightning. We recall that turbulence is motion of matter with identifiable direction at each point. Wave of any kind is turbulence and even stationary object is turbulence and when it interfaces with other turbulence such as compression or a rocket whizzing through the stationary atmosphere the dynamics of turbulence emerges.

Geological Turbulence

Our universe is turbulence from the super…super galaxy itself all the way through cosmological vortices, atoms, prima and superstrings. But, we focus on Earthly turbulence and on only five of them that fall under the standard dynamics: earthquake, volcanic activity, typhoon, tornado and lightning.

Solar Eclipse, Tidal Cycle and Earthquake

Earthquake occurs (a) when rock across fault line snaps due to uneven motion of the interfacing crust, (b) when a big chunk of the Earth's crust falls (e.g., at ocean cliff over the Philippine Deep), (c) as a result of motion of huge amount of volcanic lava underground and (d) when one of two interfacing tectonic plates 40 km below sea level pressing against each other "subducts", *i.e.*, one goes under the other causing powerful tremor and allowing lava from the Earth's interior to ooze out of the interface. Usually, the oceanic plate goes under the land or continental plate due to the impact of gravity of the Moon and Sun on the former. Every time the Moon is overhead and, especially when the Sun is also overhead we have the following configuration:

Since the Moon has the same spin as the Earth's its gravitational flux pushes the Earth's gravitational flux (by flux compatibility) and the ocean to a low tide and, in turn, pushes the tectonic plate down underneath. The Moon's gravitational flux has less effect on the porous land mass so that the effect of this action is to misalign the oceanic and continental or land plates with the former going farther down. This is reinforced by the Sun's gravitational flux when it is on the other side of the globe since it reinforces the Earth's flux on this face and pulls the ocean and plates underneath. When both the Sun and Moon are overhead (they have the same spin as Earth's) their fluxes add up and reinforce each other and push against the Earth's (by flux compatibility) pushing the ocean down to low tide and the ocean plates underneath as well. Thus, low tide at either configuration occurs every twelve hours. At noontime during New Moon and Midnight during Full Moon the push is greater and the low tide deeper because the Moon and

Sun are both overhead in the first case; the same effect, although not as much, occurs when the Moon is overhead and the Sun is on the opposite face of the Earth but the Moon's effect is greater being closer to Earth. Both configurations amplify the down push on the ocean tectonic plates. When the Sun is not aligned with the Moon on either face of Earth their combined push is the vector resultant of their individual push. In any case, the Sun's gravitational push has also less effect on the continental plate. Altogether, the ocean plates are pushed several times every 24-hr period. This regular push (greater on ocean plates) creates steady misalignment of tectonic plate interfaces. Ultimately the ocean plate subducts causing earthquake.

During solar eclipse, especially, of long duration, the rate of misalignment of the land and ocean plates rises dramatically due to perfect alignment between the Sun and Moon, advancing occurrence of earthquakes. The aftermath of the eclipse of long duration in September 1999 saw major earthquakes in Turkey (twice), Taiwan (twice), China, Iran, Indonesia, Mexico, the Philippines and Los Angeles and minor ones in the Philippines (thrice), Indonesia, Japan and Malaysia within six months. During the same period several volcanic eruptions occurred stretching from Sicily through El Salvador. Since plates are interconnected around the globe, earthquake triggers movement of plate boundaries worldwide that enhances outflow of lava at constructive plate boundaries.

Compression, grinding and lateral tension at geological faults and tectonic plate boundaries involving millions of tons of force causes interpenetration (interfacing part inducing changes on the other), collision and energetic vibration of atoms and molecules induced by the micro component of turbulence. Since this happens at the interface of turbulence the dark components of the interfacing turbulence compose and generate and propagate seismic waves. Present seismograph detects only their visible high-frequency wave components and its envelope is visible to the naked eye (e.g., the wave motion of ground surface during earthquake). Seismic wave is actually a nested fractal sequence of waves from the visible or macro through the micro and then dark component that ends up at the interface, each of the interfacing parts containing its respective fractal sequence of waves. The waves we see on the ground during earthquake are the visible envelopes of seismic waves. They are actually nested fractal sequences of waves that end up as dark (energetic) cosmic waves at the interface propagated radially from the source (recall profile of seismic wave in Chapter 3). They soften or melt metal and crack or pulverize brittle material.

The micro component of seismic waves irritate the brains of animals causing erratic behavior: horses and water buffalos jumping and running around erratically; dogs howling in distress; ants and termites coming out of their mounds and hives; roaches flying like bees to escape irritation at wall crevices and ceiling; schools of whales and jellyfish leaving the ocean depths for shallow waters; they are signs of impending earthquake. The high intensity of seismic waves just before an earthquake convert superstrings to prima that form atoms and molecules, e.g., earthlights and volatile gases like radon that shoot off flames from the ground, combined with change in water level at open wells and widening of geological faults; they are danger signs the Chinese are adept at using to predict impending earthquake.

Previously labeled UFO, earthlights of varied colors, intensity and motion have been sighted in California (along Andreas fault), Colorado, South Wales and Mexico. Geologists associate them with earthquakes but have no theory to explain them. Earthlights were sighted over Mexico City two years before the devastating earthquake of 1985. Physical characteristics of earthlights have predictive value. Bluish earthlight being energetic indicates high compression and tension at plate interface and faults and reddish one (less energetic) indicates low compression and tension. Just like the guitar string, greater tension on it produces energetic vibration and higher pitch. Then the impending earthquake can be assessed for its intensity.

Analyzing shock waves from earthquake provides information about the interior of the Earth. Since seismic waves travel faster on denser materials scientists are able to measure the specific gravity of the Earth's inner and outer core. Its core is extremely hot due to staggering spin (kinetic energy). Therefore, only prima and light elements like hydrogen exist there. They are compact which account for the great density of the core. Compression by gravity is a secondary factor. Present measurement says that the Earth's core has specific gravity of 150 [107(a)]. The present guess is that it consists of molten iron but iron is a complex atom and cannot form at such high temperature, kinetic energy and vibration. Just like the Sun the Earth's inner core (that contains the eye) consists mainly of simple prima but may also contain light elements in its outer inner core like hydrogen and helium and, certainly, its eye is also accumulating its own mini-black hole as eye of a cosmological vortex.

Along geological fault or conservative plate boundary, two kinds of dynamics stemming from independent motion of the interfacing parts and the law of uneven development occur due to the turbulent motion of the Earth's mantle under the Earth's crust. In compression and lateral tension there is grinding, interpenetration and energetic vibration of the atoms and molecules that propagate seismic waves.

This dynamics cannot be reproduced in man-made laboratory due to the huge forces involved. Verification requires simulation, observation at faults and plate boundaries and gathering data from the literature, US Geological Survey databases and even features by Sky Cable's *Discovery* (e.g., *Hostallen* Project) and *National Geographic*.

Lateral tension also involves compression due to uneven interfaces at faults (usually macro sinusoidal). Therefore, it generates seismic waves which can be monitored and level of compression and tension measured. Opposite infinitesimal lateral movement along faults is verified by macro geological motion and formation. For example, the two parts of Andreas Fault in California are moving in opposite directions that a mountain near Lake Tahoe now was in Mexico millions of years ago [107(a)]. The two parts of the Alpine Fault in New Zealand has now slid past each other by 300 km [107(a)]. Wavy fault interface adds to resistance and, therefore, power buildup; earthquake occurs when these obstacles including huge rocks between sliding parts snap and break, sending powerful shock waves (visible envelope of seismic waves) in all directions *on the ground*. However, their dark components are thrust in *all directions*. They convert dark matter to earthlights and balls of fire above the ground.

Between earthquakes, compression, grinding and lateral tension generate shock waves converting superstrings to visible matter, e.g., earth lights. They can be analyzed to study fault and plate boundary dynamics and distribution that may have practical applications.

Engineers thought that shock waves and jolting movement caused by earthquake damage high-rise structures and towers and in the 60s put rollers and springs at their base to absorb the shocks and jolting action. It did not work. The great devastation in Tokyo and Taiwan caused by earthquakes in the 80s and 90s showed that such precaution has no significant effect. From pictures of the devastation, especially, in Taiwan, the softening of metallic attachment (malleable material) at foundations and cracking and pulverization of concrete (brittle material) were quite evident. The remedy is non-existent yet: alloy that withstands the softening effect of seismic waves and composite resistant to cracking and pulverization by them.

Earthquake occurs (a) when rock across fault line snaps due to uneven motion of the interfacing plates, (b) when a big chunk of the Earth's crust falls, (c) as a result of movement of huge amount of volcanic lava underground and (d) when the oceanic plate pressing against the interfacing continental plate subducts, *i.e.*, its edge goes under the continental plate, causing powerful tremor and allowing lava from the Earth's interior to ooze out of the interface. Since the plates are 40 km underground, destruction is limited to the epicenter and its neighborhood.

We consider earthquake at geological fault, *i.e.*, upward extension of tectonic plate boundary or crack. It is irregular, usually sinusoidal, so that the relative movement of the interfacing plates causes friction between interfacing crust and tension on huge rock across it. When the friction and tension threshold are exceeded both the interfacing crust and rock across the fault snap causing tremor. Moreover, the grinding at the interface generates seismic waves that convert dark to visible matter usually in the form of balls of fire that hover along faults. As tension increases at a fault generation of seismic waves intensifies.

Tsunami

We shall focus on the devastating tsunami that occurred at the conservative interface of the tectonic plates of Burma and India off the West Cost of Sumatra, Indonesia, December 2004. The Interface goes from North to South. In this case it was the Indian plate that did subduct on that fateful day. Powerful tsunami is generated only when subduction occurs at just the right depth, about one km below the ocean surface. The subduction site extended 300 km along the interface. According to reports, during that subduction the Indian plate dropped about 50 meters while the Burma plate thrust upward some 20 meters and the whole weight of the column of water above the Indian plate dropped on the Indian plate and bounced to a height of 50 meters above the surface. Following the same water wave pattern discussed earlier, the lump of water pulled by gravity fell back on the Indian Plate at the same time the now elevated

Burma plate blocked it from going eastward, in effect, pushing it westward so that the column of water spread westward. Then the water around it pushed the spreading water back and, reinforced by the bouncing remaining column of water above the Indian plate, rose to about 50 meters above the surface again. This occurred three times pushing three crests of water waves, one after the other westward at speed of 900 km/hr to Sri Lanka, India and beyond the southern tip of India all the way to the Eastern Coast of Africa. As the tsunami enters shallow coast it slows down to 450 km/hr but like the water jet stream from constricted water hose it surges powerfully one km across the coast inland. Sumatra and Thailand being so close to the quake site only suffered flooding up to 50 meters of water without surges. The destruction was just as much as in the West due to proximity.

There was no tsunami northward and southward sparing Mayunmar (Burma) and Australia from destruction. In the aftermath of the tsunami Australian scientists reported that the earthquake vibrated the Earth, dissipated its energy and slowed down its rotation so that it will lose 1/4 day in the next century (maybe the leap year will vanish then).

Volcanic Activity

In the interfaces of compressed lava laminas or slabs moving unevenly along the Earth's crust crevices compression and grinding also occur that generate seismic waves. They convert superstrings to simple prima and photons that form coupled prima, atoms and molecules that produce earthlights and balls of fire (earthlights high up in the atmosphere, in the mesosphere, while balls of fire hover just above ground or ocean surface). Their physical characteristics shed light on accumulation of lava long before they reach ground level. They can be used for calculating the power of an impending eruption and predicting its occurrence. Like visible components of shock waves from seismic activity they are detected as high-frequency waves by seismographs. In fact, huge movement of lava causes slight tremors also that may generate seismic waves around the crevices due to the motion of huge rocks and crust. The fractal structure of seismic wave is reflected in the structure of the lava outflow. Some geologists study it to get underground information. Technology can be generated to measure and monitor lava accumulation and movement from tectonic plate boundary through the Earth's surface and gather information revealed by lava. For instance, most living organisms come from volcanic lava. Thus, the volcanic Galapagos Islands off the western Coast of Ecuador have the most complete collection of animal species. Darwin spent much time there in his study of the evolution of species. Perhaps, the volcanic Hawaiian Islands come second to Galapagos in diversity of animal species. Ocean species also abound in the boiling waters of the deep around under-ocean volcanoes. (This information tells us that ingredients of living organisms also originate in the interior of the Earth)

Seismic waves (fractal seismic waves from macro to dark frequency) from volcanic activity are generated as follows:

(1)   Grinding of lava slabs under extreme temperature,

(2)   Motion of boulders that give way to lava flow and accumulation that induce compression, lateral tension and grinding at faults and

(3)   Effect of lava flow as it passes across tectonic plate boundaries and crevices.

Most lava flows out of constructive boundary (constructive because the lava piles up into mountains on both sides of the boundary). It induces separation and, therefore, intensifies compression at the opposite subductive boundary of the plate where earthquake occurs. Thus, there is a close link at different levels between volcanic and seismic activity. Its impact on visible matter is verified in the Philippines by earthlights and vigorous cloud motion over seemingly dormant volcanoes. Moreover, the mysterious under-ocean bright lights in the Southern Philippines several years ago may have link to the strong tremors that followed (they could have been under-ocean earthlights).

Areas of intense volcanic activity lie along constructive and subductive plate boundaries; constructive because they are open and allow lava outflow that deposit huge piles of lava. The Pacific Ring of Fire (string of volcanoes) along the Pacific Rim consisting of constructive, subductive and conservative plate boundaries generates intense volcanic activity. They have direct link with *el niño* and, ultimately, cyclones. There is minimal volcanic activity at conservative boundary, only when there is subduction. Along constructive plate boundaries massive outflow of lava heats up the ocean surface to form *el niño* but also accumulate along plate margins that materially affects atmospheric behavior and volcanic activity.

Atmospheric Turbulence

Atmospheric turbulence includes tropical cyclones and hurricanes, tornadoes and lightning.

The Pacific and Atlantic Wind Cycles

Denser material offers more resistance or viscosity to the Earth's gravitational flux. Therefore, it is more efficiently pulled with minimal slip or lag by it. The Earth's hot compact inner core of pure prima around the event horizon of the Earth's eye has specific gravity of 150 like the Sun's deep interior. Therefore, it is most efficiently pulled by the Earth's gravitational flux (from West to East; this is the dual of the vortex flux of superstrings induced by the toroidal flux of the protons in the nucleus, circular speed: 7×10^{22} cm/sec [4]) and spins at staggering rate. Moving outward from the core, density declines and the material slips more relative to the inner material. Therefore, the less dense waters of the Oceans, particularly, the Pacific Ocean, lag behind the Earth's denser crust at the Equator. The effect is ocean current from East to West starting off the Peruvian Coast towards the Asian mainland. Blocked by the land mass in Asia, the northern slice of the ocean flow veers northward (the southern slice veers southward) towards the North Pole at zero gravitational flux; however, blocked by the Chinese landmass it veers Northeastward through the Sea of Japan, then Eastward South of the Bearing Sea, following the contours of the Asian Mainland, and down the Eastern Seaboard of Canada and the US towards West of the Peruvian Coast to complete the Northern Pacific Ocean Cycle. The layer lag along the segment from the Korean Coast to Alaska is neutralized by the polar lag so that the wind cycle encounters no resistance there. Ancient traders used the Northern Pacific Wind Cycle to facilitate travel, especially, off the Asian mainland through and the Western seaboard of North America.

Along the route there are eddies or local elliptical cycles of compatible spin with the Northern Pacific Ocean Cycle. Local fishermen use them to facilitate travel also.

There is a similar lag between the lower atmosphere and the Pacific Ocean that translates into the Trade Winds along the Equator starting west of the Peruvian Coast going westward. However there is a difference: instead of being blocked by the Asian Coastline, the northern slice crosses into South Vietnam; blocked by the mountainous interior of the Asian Mainland, it veers north-eastward grating North Vietnam and China and off across the Sea of Japan through South of Siberia across the Bearing Sea, crosses the Western Seaboard of Canada and veers right into the Tornado Belt of the US between the Midwest and Texas and between Oklahoma and Florida. Then the wind flow goes down through it, veers to the right, crosses the Mexican Coast and joins the Trade Winds to complete the Northern Pacific Wind Cycle. (Part of it is the jet stream from the Sea of Japan through south of Siberia that airlines utilize to save fuel going East). Typhoons that originate in the Equator East of the Marianas Trench (deep interface of tectonic plates that spew off volcanic lava) follow the course of the Northern Pacific Wind Cycle.

The Northern Pacific Wind Cycle has also local elliptical cycles that serve as routes for local traders and fishermen, especially, along the Eastern Seaboard of Asia.

The Northern Pacific Wind Cycle interfaces at the Tornado Belt with a similar Northern Atlantic Wind Cycle coming from the Caribbean through the Gulf of Mexico, veers north and grates the Tornado Belt, northeastward into Eastern Canada, off East to the Atlantic Ocean, veers east-southward and grates Western Europe down to the Western Mediterranean and veers right to the Caribbean to complete the cycle. The Northern Pacific Wind Cycle determines paths of typhoons in the Western Pacific and, together with the Northern Atlantic Wind Cycle, occurrences and paths of tornadoes in the US Tornado Belt. The interface of the narrower Northern Atlantic cycle and Northern Pacific cycle determines the paths of tornadoes through it over the Tornado Belt. The interface oscillates between Oklahoma and Florida and scrapes the Tornado Belt.

(The Northern Pacific cycle has a counterpart in the Southern Hemisphere but it is not as coherent and strong because of the broken land masses there; only baby tornadoes and sprouts occur)

Typhoon and Tornado: Eddies in Earth's Gravitational Flux

Winding bank and bank irregularity make water in spring and rapids flow unevenly. Imagine an interface of current moving forward, the right current faster than the left. Represent the right current by a longer arrow than the arrow representing the left. As the interfacing currents move forward, the left current pulls or drags a slice of the right current at the interface to veer steadily leftward, wraps around a slice of the left current that also veers leftward steadily and joins and resonates with the other slice of the right current to form a vortex, an eddy. The eddy serves as ball bearing that minimizes friction and dissipation of energy. The right rim of the eddy is reinforcing since it coheres and resonates with the other slice of the right current and the left rim repels the other slice of the left current (by flux compatibility) opening a thin gap between them that minimizes friction and dissipation of energy. Water eddies are abundant on the downside of rapids; it is interesting to observe their direction of movement and realize that it is subject to natural law, particularly, energy conservation.

The equatorial corridor off the Peruvian Coast all the way through the Marianas Trench in the Western Pacific is studded with shallow under-ocean volcanoes oozing with hot lava through constructive plate boundaries and cracks [107(b)]. Volcanoes have their cycles. During periods of volcanic activity hot lava heats up the ocean surface and forms *el niño* that, in turn, heats up the lower atmosphere. When the heated ocean surface is contiguous and huge, say, as broad as the Canadian landscape, it creates a huge tropical depression in the lower atmosphere above it that sucks the surrounding air molecules by the trillions, by flux-low-pressure complementarity. Initially chaotic due to collision of such huge number of molecules, energy conservation induces coherence of fluxes that results into coherent vortex flux of air molecules called tropical cyclone (typhoon or hurricane).

Why does warm air create low pressure or depression? It infuses kinetic energy on the gas molecules, intensifies Brownian motion, raises collision intensity and pushes the gas molecules against each other (pressure is in all directions). It disperses the gas molecules, reduces the density of the air and creates a region of low pressure or atmospheric depression. Low pressure pushes air upwards over the depression and sucks air molecules around it.

In the Northern Hemisphere there is a lag in linear speed of the gravitational flux (from West to East called polar lag) from the most intense at the equator to 0 in the North Pole. Like the water current this linear flux intensity gradient gives a counterclockwise twitch to the rising gas molecules of developing cyclone through resonance with their dark component, to spin counterclockwise. This is the reason all typhoons originating in the Northern Hemisphere have counterclockwise spin. There are rare cases where a typhoon originating in the Southern Hemisphere crosses northward across the Equator and retains its clockwise spin due to spin momentum already acquired.

In the Western Pacific typhoons originate East of and around the Marianas Trench. They follow the contours of the Northern Pacific Wind Cycle unless distorted by temperature variation over the Philippine Deep where there is a string of volcanoes at the bottom (part of the Pacific Ring of Fire) and around Mayon Volcano in the southern tip of the island of Luzon. Typhoon in progress goes along warm corridor that enhances its power, by energy conservation.

Energy conservation forms and shapes the eye of a typhoon and effectively minimizes friction of air passing through it. The faster the wind spin the bigger the eye due to centrifugal force that pushes the "eye wall" outward. Therefore, when typhoon passes over a mountain and the eye is shut off and friction weakens it. This is the reason typhoons that hit Central Luzon and Metro Manila are generally weak due to the Mountainous corridor along the Eastern Coast of the country. However, when the eye of a typhoon passes across gaps in the mountain ranges or the sea lane south of Luzon and veers northward across Metro Manila the latter may be hit by a big one.

Being too deeply located the volcanoes in the Philippine Deep only give rise to broken *el niño* rather than huge solid compact *el niño* that creates tropical depression and gives rise to typhoon. Consequently, they only influence its course and create chaotic wind movement that hampers the normal evaporation of ocean water the effect of which is to bring drought to this region. That is not the case in the Eastern Pacific where hurricanes are formed instead of chaotic wind motion.

At the Tornado Belt in the US the Northern Atlantic Wind Cycle brings warm air from the Equator while the Northern Pacific Wind Cycle brings cool air from the Northern polar region (by flux-low-pressure complementarity) during spring and early summer. With the denser cool air from the North and the lighter warm air from the Equator pressing against each other over the Tornado Belt, the former goes down and the latter goes up forming a horizontal cylindrical vortex initially from North to South along the southern segment of the interface and North to Northeast along the upper half. This happens about 1.5 km above sea level. By resonance with the gravitational flux polar lag in the Northern Hemisphere, the Northern end of the vortex tilts downward and spins counterclockwise as it sucks the denser air towards the ground and by Newton's action reaction law that end swoops down and must touch down and press firmly on the ground by suction to form a tornado successfully. Thus, the tornado funnel spins counterclockwise in the Northern Hemisphere. A tornado in the Southern Hemisphere is the mirror image of one in the Northern Hemisphere with respect to the Equator. The most powerful tornadoes to hit the Belt (Oklahoma, 1995; Texas, 1998) had 510-km spin around the 91-meter-diameter eye (intensity F-5). The eye is larger the faster the spin, by centrifugal force, and the greater the suction. Suction is weakened by rugged terrain on the ground, slope or thick forest. The dark funnel visible in the Oklahoma tornado was only the eye and its immediate vicinity; the whole funnel was one kilometer across that cleaned up a strip six km long and one km wide.

The spinning funnel of the tornado in the Tornado Belt is like a horizontal wheel that spins counterclockwise and since left of the interface (cold wind from the North pulled by the Northern Pacific Wind Cycle, by flux-low-pressure complementarity) is denser, there is greater friction there than on the right of the interface which is warm and lighter coming from the warm Caribbean. This propels the funnel to move along the interface from the South sector direction towards the North in the southern portion of the Belt and slightly tilted towards the East in the northern portion due to the smaller but more forceful Northern Atlantic Wind Cycle at speed of about 30 km/hr.

As the tornado travels, it grates the ground, flattens structures and its eye sucks debris upwards and scatters it within a 210-km radius, light debris like paper and cardboard thrown farthest. The macro envelope of the micro component of turbulence scoops up logs, poles and lumber from the ground and the tornado spin of the funnel catapults them in the forward direction (called the torpedo effect) and hit structures and weaken them. The funnel flattens them and the eye sucks and throws them around. Both the Oklahoma and Texas tornadoes cleaned up a lane six km long and one km wide, each eye as big as a football field.

Typhoon

We look at typhoon in greater detail. We first remark that vortex fluxes, visible or dark, is subject to the same two laws of nature, flux compatibility and flux-low-pressure complementarity. The reason is: pressure is a collective phenomenon independent of the individual properties or structures of the physical systems that form the flux. Thus, the spin of typhoon or tornado is the same as the spin of its dark component and subject to the same laws that govern the behavior of water eddy. In the northern hemisphere, for instance, the polar lag in the Earth's gravitational flux from the Equator to the North Pole causes the dark component of the typhoon or tornado, as dark eddy, to spin counterclockwise. This is the reason for of the counterclockwise spin of typhoon or tornado in the Northern hemisphere.

Typhoons form over shallow constructive plate boundaries along the Pacific Ring of Fire and cracks in oceanic plates that form under-ocean volcanoes and hot spots. The Pacific Ring of Fire has branches consisting of plate boundaries most of which constructive. Constructive plate boundaries too deep in the ocean such as west of the Marianas Trench and the Philippine Trench at the bottom of the Philippine Deep do not contribute to formation of typhoons but influence their paths. The *el niño it* forms are not contiguous but broken so that they give rise not to coherent vortex flux of air but incoherent and chaotic air wind movements that hamper the normal evaporation of water that produces rainfall. They cause drought rather than typhoon.

The region east of the Marianas Trench through the Central Pacific and on to the Coast of Central America down to the Peruvian Coast is studded with volcanoes and, in the Central Pacific and around Galapagos Islands, also hot spots. Typhoons originate there. The region supplies contiguous *el niños* that create and power typhoons.

First, there is calm over these regions, say, in the Northern Hemisphere, the initial phase of the standard dynamics. Then huge solid *el niño* pocket forms, sometimes as broad as the Canadian landscape and as thick as 650 meters deep. This warms up the air and raises kinetic energy of the air molecules, pushing each other away to thin out the lower atmosphere, induce low pressure and cause air to rush out through thin air above and form atmospheric depression. By flux-low-pressure complementarity, depression sucks denser air around and causes initially chaotic inward rush of air due to collision; this is the second or transitional phase of chaos. Chaos is energy dissipating and, by energy conservation, stabilizes into coherent flux called turbulence, specifically, a vortex called cyclone (typhoon or hurricane), third phase of the standard dynamics. As we have seen earlier a vortex spins counterclockwise in the North; this is due to the counterclockwise twitch caused by the polar gravitational flux lag towards the North Pole that resonates with the dark component of the rising column of air (exactly like the way water eddy is formed when there is uneven water flow or current; the tangential component of the gravitational flux is most intense at the Equator and goes to zero at the Poles). This results in helical streamlines (paths of air molecules) from the ocean surface through the rising column around the vertical cylindrical eye with micro local oscillation as local component of turbulence. As it leaves the eye the helical path is thrust outward by centrifugal force but its vertical profile reaches its peak along parabolic arc at the vertex and goes downward also along parabolic arc, in accordance with Newton's mathematical model of an object thrown into the atmosphere subject to the pull of gravity, with vertex somewhere midway between the boundary of the eye and the rim.

Then the helical path reaches the bottom and winds back to (and is sucked by) the eye, winds around upward again and the cycle repeats, etc. The vortex is donut-shaped but flattens at the bottom as it grates the ocean. The vortex is covered by these helical streamlines; its top having parabolic profile but almost flat due to centrifugal force on the molecules imparted by vortex spin that stretches it.

The vortex is almost solid; *almost*, because there is a secondary horizontal circular cylindrical eye at the middle of the donut shaped vortex along its axis and around the main vertical eye. This secondary eye is verified during typhoon by the circular motion of tree branches as this eye passes by. Thus, a typhoon is donut-shaped vortex with two eyes the main one being the hole of the donut and the other called axial eye, a hollow cylindrical eye along the donut's circular axis. Here, the typhoon is an eddy in air current and its dark component an eddy in the Earth's gravitational flux. As long as there is upward flux the counterclockwise twitch sustains the spin and centrifugal force and momentum imparted on the helical flow of molecules. This initially increases the typhoon's power until it reaches its peak. Then the vortex begins to ease the depression and erode its own power and the depression fades out as soon as the vortex resolves the pressure gradient with the surrounding atmosphere through interpenetration.

The eye minimizes collision and friction, *i.e.*, dissipation of energy. Eye formation is forced by centrifugal force of spiraling molecules that push the other molecules outward. Thus, the greater the spin, the bigger the eye is. When the typhoon's main eye is plagued off by a mountain along its path the cyclone loses steam. This explains weakened state of typhoon over Manila that passes first over the mountain ranges Banahaw east of the city. Only a few times when a typhoon detours around the mountain ranges of Quezon and Laguna Provinces into the sea then turns northward to Manila does it hit the city hard. The typhoon has the effect of leveling up depression and weakening its own power. As it weakens, centrifugal force on spinning air molecules declines, the main eye shrinks and the typhoon reduces in size and fades away; or, the vortex may simply break and become incoherent wind movements followed by calm.

There is substantial verification of typhoon dynamics aside from spin. Rural folks know when typhoon approaches and distinguishes it from ordinary monsoon wind: in an approaching typhoon broken tree branches shoot downwards and pierce the ground because that is the direction of the streamlines. In ordinary monsoon wind branches just fly off. They also distinguish the calm when the typhoon's eye is passing by from the calm when it is over as the wind direction reverses after the eye has gone by. They know where the path of the eye is relative to their location by the manner by which wind direction shifts. In the Western Pacific typhoons normally follow the contours of the Northern Pacific Wind Cycle.

However, frequent heating up of the ocean surface above the Trench at the bottom of the Philippine Deep that yields broken *el niño,* while not sufficient to form tropical depression, combined with temperature variation around Mayon Volcano affects the path of an approaching typhoon: it moves towards hot corridor with less atmospheric pressure, pulling southward instead of following the contour of the Northern Pacific Wind Cycle. Then its path can be predicted by monitoring departure of its path from normal course along the Northern Pacific Wind Cycle as it crosses into the Philippines area of responsibility west of the Marianas.

It is theoretically possible to reroute a typhoon but that would be impractical. It is more advantageous to cope with and take advantage of it by utilizing its energy, e.g., floods. The flood waters can be stored and recycled for purposes of hydroelectric generation, irrigation and treatment for potable water.

Ref. [107(b)] provides data on the relationship between under-ocean activity and hot spots and *el niño* and, therefore, topical depression, cyclones and tornadoes. With this theory we can now explain the difference in impact of *el niño* between the Eastern, Central and Near Western Pacific regions (up to the Marianas Trench), on the one hand, and Western and Southwestern Pacific, on the other. Volcanic activity abounds in the Central Pacific from the Hawaiian Island group through Marshall Islands and the Marianas where most tropical depression forms. There is moderate volcanic activity along trenches around the Philippine Plate, having mainly destructive boundary there; the 15% conservative plate boundary there hardly contributes to volcanic activity. Moderate volcanic activity along the trench in the Philippine Deep does not produce solid *el niño* but only broken little pieces that induce chaotic rush of air molecules that stabilize as local incoherent fluxes that interfere with normal evaporation of water but does not stabilize into cyclones. They do not sufficiently heat up air above the ocean. This explains frequent drought in this region. However, they reduce pressure sufficiently to extend the typhoon corridor toward the China Sea. This theory provides cheap and accurate means of predicting a typhoon's course without radars that get knocked off by advanced strong winds anyway.

On the other face of Earth are numerous under-ocean volcanoes in the Atlantic just East of the US and North of Cuba and across the Caribbean aside from the constructive plate boundary across the Equator from North to South between the Continental Americas and Europe – generators of devastating hurricanes in that region. There is also constructive plate boundary just west of the US from Oregon through Washington State and Canada and another one west of Central and South America, aside from the group of volcanoes and hot spots around Galapagos Islands that extend far towards the Central Pacific; they are also generators of huge *el niño* and devastating hurricanes

Tornado

Tornado occurs at sharp interface of hot and cold air pressing against each other. In the US the Tornado Belt where most tornadoes hit is the 1000- by 650-km corridor between Oklahoma and Florida and between the Midwest and Texas. This is where the Northern Pacific and Atlantic Wind Cycles interface and press against each other, by flux compatibility, the former bringing cold air from the North, by flux-low-pressure complementarity, the latter warm air from the Equator. Since the Pacific Wind Cycle, which is roughly elliptical, is much wider than the Atlantic Cycle, the interface over the Tornado Belt goes from North to South but slightly tilted towards the East in the North. It oscillates between Oklahoma and Florida. The cold air from the North merges with the Pacific Wind Cycle in this interface. The temperature gap occurs in spring and early summer, but the gap is greatest in early summer when strong tornadoes occur.

At the interface, warm air from the Equator presses against cold air from the North and goes up, being lighter, while the denser cold air goes down. This action creates a little vortex with horizontal axis going in the general direction from South to North. The gravitational flux twitch combined with the greater force of suction on denser air tilts the north end of the vortex downwards by flux compatibility with the Earth's gravitational flux. Since the denser air comes from the bottom end of the vertical eye it pulls the vortex down, by Newton's action-reaction law, and swoops towards and touches down on and grips the ground by its suction. At this time we have a full-blown tornado with a counterclockwise spin of its funnel.

The suction or grip is firm and stable on flat ground but on rugged surface or incline it is jumpy, does not stabilize, lifts off and breaks up, the reason powerful tornadoes are only stable on the plains. Thickly forested areas also weaken its power.

The counterclockwise spinning funnel is like a wheel. Since the denser air offers more friction than lighter air the spinning funnel is pushed northward with easterly tilt greater in the northern section of the Belt due to the broader Northern Pacific Wind Cycle. Thus, a tornado in this Belt travels northward (with easterly tilt) at about 30/kph. As the tornado moves forward several things happen: (a) it flattens structures and grates the ground; (b) in the interior of the funnel the debris is sucked by the eye, pushed up and scattered outward by as much as 210 km, farthest thrown being light material like paper and card; (b) near the edge of the funnel on the ground the micro component of turbulence scoops up material, e.g., log, post, lumber from debris, and, by the combination of centrifugal force and forward push of the funnel, throws them as missiles (this is called torpedo effect) at structures along its forward path and weakens them before the funnel flattens them. The funnel remains strong as long as there is suction and sufficient temperature gradient between the two interfaces.

Tornado touchdown is generally of short duration due to interpenetration that wipes out the sharp interface of warm and cold air. Its duration ranges from split second to a couple of hours. The tornado of longest known duration lasted for over a couple of hours and traveled in the northeasterly direction (typical in the North end of the Belt) from Monroe City, Wisconsin.

Sharp interface of warm and cold air can happen anywhere because of the irregular course of wind flow but it is rare and not as powerful outside the Tornado Belt. However, when it happens at sea it forms waterspout that scoops up water and dumps it elsewhere, sometimes on land. This can create a deep hole on the ground or damage ships and harbors. Weak tornadoes also occur in Australia due to the interface of the Southern Pacific Wind Cycle and the narrow and the weak and narrow wind cycle over the Indian Ocean.

Clearly, this section has solved the turbulence problem for these four types of turbulence and opens the door wide open for their control (partially for typhoon, earthquake and volcanic activity but full for tornado); a new technology has been conceptualized for monitoring potential tornado occurrence and aborting, deflecting and terminating or breaking a tornado in progress to protect cities and farms.

Lightning

Lightning occurs when there is accumulation of positive ions in the lower atmosphere and the electrical potential between the ions and the ground reaches critical level. The electrical potential depends on the amount of positive ions since there is abundance of electrons on the ground due to the push by the Earth's gravitational flux. When critical potential is reached, the positive ions rush down and the negative ions rush up. They collide resulting in explosion. Some estimates put the energy of a bolt of lightning at one megaton of thermonuclear explosion.

Since lightning is an explosion it generates and propagates seismic waves that convert dark matter to earthlights in the mesosphere, some 50 to 90 km above the Earth's surface. They are familiar spectacle in the skies for airline pilots.

How do positive ions accumulate in the lower atmosphere? One source is the gravitational push by the Earth's gravitational flux, by flux compatibility. Another is turbulence that knocks off valence electrons of gas and turns them into positive ions. This occurs, for instance when there is forming cyclone and tornado in the clouds and even when the tornado has touched down the spinning funnel continues to ionize the air close to the ground in which case the critical electrical potential is low. Thus, a tornado in progress has its funnel wrapped in little sparks of lightning close to the ground. This also occurs, especially, during the chaotic phase in the formation of typhoon. However, lightning stops when the electrical potential has been wiped out by the release of energy.

THE SOLUTION OF THE GRAVITATIONAL N-BODY PROBLEM

This problem has a unique place in natural science for it catalyzed the development of GUT then called the flux theory of gravitation [33, 41]. Its solution required the discovery of the superstring, basic constituent of matter. The crucial factor in the solution was the introduction of the new methodology of qualitative modeling.

Historical Background

Posed by Simon Marquis de Laplace at the turn of the 18th Century in his book, *Celestial Mechanics*, in his attempt to prove the stability of the solar system, then thought to be our universe, the gravitational n-body problem raised some fundamental questions that sparked the development of GUT. Its solution was the first milestone in the historical and theoretical development of GUT.

In 1992 this question was posed: why are there long-standing unsolved problems of mathematics and physics that defied resolution for so long? For both mathematics and physics the answer is: inadequacy of their underlying fields. For physics the principal underlying field is mathematical physics and the inadequacy lies in its methodology of quantitative modeling that only describes appearances of nature mathematically and statistically. The problem says:

Given n bodies in the Cosmos at time T and positions, $x, ..., x_n$, velocities $v_1, ..., v_1$, and of masses $m_1, ..., m_n$, subject to their gravitational interaction, find their positions, velocities and paths or trajectories at later time.

Something is missing here: Neither "body" nor "gravity" is known. Therefore, we do not know the behavior of this physical system and cannot predict its future course required by the problem. Even the "solutions" of the simplest case, n = 2, are not satisfactory. One solution presented at the 2nd International Conference on Dynamic Systems and Applications, May 1995, Atlanta, says that the Earth orbits around the Sun along a rotating plane whose axis of rotation coincides with the Sun's polar axis which is not true. It requires a physical theory based on some laws of nature, to define the relevant concepts involved and solve the problem. Newton's so called law of gravitation is not a theory but a quantitative model, *i.e.*, description of motion of bodies under the influence of that unknown force called gravity. Even "body" is not defined properly. To know what a body is it is necessary to answer what the Greeks and the Chinese and, inevitably, most cultures, sought several millennia ago: what the basic constituent of matter is. Physicists have been smashing the nucleus of the atom for over 50 years to find it. They failed because it is not the right place to find it. The irony is the electron was discovered by J. J. Thomson in the early 19th Century and it is an agitated superstring.

The characterization of undecidable propositions in the course of the resolution of FLT revealed the inadequacy of quantitative modeling: mathematical and physical spaces are distinct, therefore, independent; the former being man-made and defined by its axioms, the latter objective and subject to and defined by the laws of nature. Therefore, a problem in one space cannot be solved by the other; one can only use reasoning by analogy; sometimes it works but not all the time. This is the context by which we can look at the inability to solve the gravitational n-body problem for two hundred years.

The solution of this problem in 1997 [41] required the discovery of 11 laws of nature; they were the initial laws of nature of GUT required to discover and define the basic constituent of matter, the superstring, and its conversion to visible matter. They also define gravity in its full manifestation as a natural phenomenon, *i.e.*, as dynamics of a cosmological vortex. In Laplace formulation it was expected that the solution would yield the n bodies forming a planetary system like Jupiter and its moons where one body, Jupiter, is at the center and the others orbiting around it or, like the planets, would find their respective orbits around the Sun.

It is not quite that simple now because the bodies may belong to or be under the gravitational influence of different cosmological vortices. Moreover, the solution depends on what kind of bodies they are. Let us then classify the bodies in the solar system first and then generalize the solution beyond the solar system.

(a) Cosmological body, *i.e.*, the core or collected mass around and at the event horizon of the eye of a cosmological vortex to which category belong the stars, planets, planetoids and moons.

(b) Comet, a damaged cosmological body, e.g., planetoid, due to near hit on the Sun but damaged and catapulted into an elongated orbit by the Sun's gravity (we confine our definition to the solar system because we do not know if a comet has counterpart in a galaxy).

(c) Debris, e.g., asteroid, comes mainly from comet's tail or planetary collision with planetoid or collision among wayward planetoids; they cluster around huge planets like Jupiter and Neptune because their strong gravity serves as trap for collision except those that get suspended in neutral regions (*i.e.*, between planetary vortices and under the Sun's dominant gravitational flux influence).

(d) Meteors, debris from tail of comet light enough to be attracted by or plunge into weaker planets like Earth.

(e) Cosmic dust, converted prima from dark matter that gets entangled into micro cosmological vortices. Technically, cosmic dust consists of cosmological vortices so light they cannot move much as their dark component cannot overcome dark viscosity (their motion is analogous to Brownian movement but confined to narrow neighborhood). One thousandth of Milky Way's mass is cosmic dust [4]. Of course, the bulk of its mass is generated by the micro component of turbulence at its inner core. Items (b) to (e) were already discussed earlier but we shall summarize the discussion here and explain their behavior. Moreover, in the case of (a), the qualitative solution of the n-body problem was also done under macro gravity. That is exactly what the 11 initial laws of nature did: provide the qualitative solution. The bodies lie along spiral streamlines the latter rotating around the core; for some, when centrifugal force equals the pull of gravity, they take elliptical orbits. Then the integrated Pontrjagin maximum principle provided the computational component of the solution that we shall undertake here.

This is really the role of computation in physics: to corroborate, concretize and provide details of the qualitative solution. The computational tool we use is the Pontrjagin maximum principle, specifically, to find the equations of the trajectories of the n-bodies. We shall take the general case were the n bodies do not all belong to the same cosmological vortex. In view of the fractal nature of our universe with itself as the common first term of the fractal sequences [65] there is a minimal cosmological vortex that contains all of them. We put the origin of our global coordinate system at its eye. Then with the fractal-reverse-fractal locator [36] we locate the positions of the n bodies and using suitable composite functions find their equations relative to the global coordinate system.

The Solution of the Gravitational n-Body Problem

We first solve the problem as a time-optimal problem on a cosmological body by formulating it in the equatorial plane of the core vortex where the problem is to bring the body (a minor cosmological body) to the surface of the collected mass at the main eye where the origin of the local coordinate system is located at its center. This is an uncontrolled problem in the context of the Pontrjagin maximum principle because the controls are fixed: the pull of gravity and the centrifugal force on the body. This makes the solution a relaxed trajectory, a special case of generalized curve, having set-valued derivative. The control set consists of the values of the derivative constrained by some differential equation. More set-values of the derivative of a relaxed trajectory come from the micro component of turbulence along the streamlines as interfaces of turbulence.

We consider the family,

$$H = yG, \tag{1}$$

of Hamiltonian functions $h(t, x, y) = yg(t, x)$, where y is a variable vector and each $g \in G$ gives rise to a corresponding $h \in H$. We shall be concerned with points (t, x) that lie in a sufficiently fine neighborhood of the set described by a given fixed trajectory C of the form $x(t)$, $t_1 \le t \le t_2$. In terming C a trajectory we imply that the function $x(t)$ is, almost everywhere in its interval, a solution of the differential equation,

$$\dot{x} = g(t, x(t)), \tag{2}$$

for some fixed corresponding member $g \in G$; moreover, $x(t)$ is to be absolutely continuous. Here G is a convex family of functions, a family such that every convex combination,

$$\Sigma\alpha_i g_i, \tag{3}$$

of a finite number of members g_i, of G with constant coefficients $\alpha_i \ge 0$, where $\Sigma\alpha_i = 1$, is itself a member of G. In addition we require that every function $g(t, x)$ in G is continuously differentiable in x for fixed t and measurable in t for fixed x, and also that each g and its partial derivative g_x are bounded functions of (t, x) or, more generally, bounded in absolute value by some integrable function of t only. These various requirements are to hold only in some bounded open set O of (t, x) space. In the chattering case all these requirements are satisfied if we make the stipulation that $g(t, x, u)$ is continuously differentiable and strengthen the convexity condition by requiring that the α_is are measurable functions of t.

(In the formulation of the Pontrjagin maximum principle x is a vector in n-space; so that the optimal solution of (2) would be an absolutely continuous vector function in n-space. Here the solution will be an absolutely continuous parametric function in the plane in the parameter t)

The appropriate case for this problem is the matrix A having eigenvalues with negative real parts. Then trajectories $x(t)$ approach the origin for large t, *i.e.*, the system is stable. By appropriate choice of unit of time (change of scale) we assume the eigenvalues to be $\lambda \pm i$, or $\lambda, \lambda - 1$, where $\lambda < 0$. (Ref. [117] uses i simply as notational convenience suggesting awareness of the problem with i) By affine transformation of x and translation of u we suppose that B is unit matrix and take the appropriate case where A is a 2×2 matrix whose rows are given by:

Row 1 = $\lceil \lambda -1 \rceil$
Row 2 = $\lfloor 1 -\lambda \rfloor$. $\tag{4}$

To find the optimal control function we use Young's theorem [120].

Theorem

Let u(t) be a suspected optimal control which transfers our particle from the initial position x_0 at time t_0 to the origin at time t = 0 and suppose that the corresponding function x(t), which defines the trajectory described by the particle is not constant in any time interval terminating at t = 0, i.e., that the final arc of the trajectory does not reduce to a single point, consisting of the origin. Then u(t) is an optimal control, and the only one to make this transfer.

The control parameter u belongs to a convex set U, a parallelogram none of whose edges is parallel to a coordinate axis. The Pontrjagin maximum principle asserts the existence of nonvanishing solution $y = \psi(t)$ of the adjoint equation,

$$dy/dt = -A^*y, \tag{5}$$

where A^* is the transpose of (4), such that the Hamiltonian (scalar product),

$$H(x, y, u) = (y, Ax+Bu), \tag{6}$$

attains, when $x = x(t)$, $y = \psi(t)$, its maximum in $u = u(t)$. The initial condition at time t_0 is $x(t_0) = x_0$ and at final time at the origin both $x = 0$ and transversality condition in t which states that $H > 0$ for the argument $(x, y, u) = (0, \psi(0), u(0))$ holds. Ref. [117] shows that the condition $H > 0$ holds on suspected solution and is constant there and hence can be verified at any time t along that trajectory. The control u(t) is piece-wise constant.

We term switching time for suspected solution one for which u(t) is discontinuous in t. For fixed t the maximum of $H(x(t), \psi(t))$ is attained when the scalar product $(\psi(t), Bu)$ is maximum and, being linear, attained at a vertex of U. It occurs there only unless there is one edge with direction vector w on which the scalar product is constant so that by subtracting w it satisfies,

$$(\psi(t), Bw) = 0. \tag{7}$$

The solution of the adjoint equation is,

$$y = ce^{-\lambda t}\eta, \text{ where } \eta = e^{j(t+\alpha)}, \tag{8}$$

where α and c are integration constants and $c > 0$.

Note that the η-curve is a rotating unit circle in the counterclockwise direction so that the product with the radial curve $y = ce^{-\lambda t}$ tends towards the origin as t increases, *i.e.*, the spirals tend towards the origin. We can interpret this as the path of a falling body in a cosmological vortex with counterclockwise spin. This family of spiral curves covers the whole vortex. Of course, the mirror image of this family relative to the y-axis is a family of spirals in the clockwise direction and corresponds to the paths of falling bodies in a cosmological vortex with clockwise spin.

If there is another body in the same cosmological vortex then there is a unique spiral that connects it with the origin. It does not really matter if the body is in orbit or not, the body's trajectory will be a rotating spiral.

For each body in this cosmological vortex we have all the information the problem asks for: position, trajectory and velocity.

General Case of the Problem

We have solved the uncontrolled problem for one body in a cosmological vortex. When there are more than one then each body lies in some spiral and we simply find their equations one at a time.

When the n bodies do not all belong to the same cosmological vortex then, in view of the fractal nature of our universe with itself as the common first term of its nested fractal sequences, there exists a (minimal) cosmological vortex that contains the cosmological vortices of the other bodies of the problem. We place the global coordinate system in that vortex with the origin at its center. (The common cosmological vortex of the n bodies can be our universe itself) To find the locations of the other bodies we apply the fractal-reverse-fractal vortex locator on each [36]. Then we find their equations with respect to the global coordinate system. We do it one at a time for each body.

Without loss of generality we take the Moon as one of the n bodies. Using (8) we can find its rotating spiral trajectory about the Earth. The Earth may be one of the bodies, it does not matter. At any rate, we can find the Earth's rotating trajectory in the local coordinate system with origin at the center of the Sun. Then we can find the appropriate composite function that gives the equation of the rotating trajectory of the Moon in terms of the local coordinate system with origin at the center of the Sun.

Moving backwards, we can find in exactly the same manner the equation of the rotating trajectory of the Sun about the center of the Milky Way in terms of the local coordinate system with origin at the center of the Milky Way and the appropriate composite function that would yield the equation of the Moon.

We can do the same thing for all the intervening cosmological vortices from the Earth through the common cosmological vortex and give the equation of the Moon in terms of the global coordinate system located there. Then we can replicate this scheme for all the other bodies of problem.

Solution for the Other Categories of Bodies in the Cosmos

Massive piece of debris in the cosmos, e.g., asteroid, comet, has greater viscosity relative to the gravitational flux it is under the influence of. Long before an asteroid reaches the vicinity of a cosmological vortex, say Earth, it gets deflected away from it by the gravitational flux. The Earth's gravitational flux reaches far beyond the Moon and an asteroid starts getting deflected there. Only when it enters the narrow injection angle of 2° (used as guide by astronauts in coming back to Earth) will it crash into Earth. Even when it intersects the spiral streamline in the same direction, its momentum is too strong to be sucked by the Earth's gravitational pull towards the eye along that streamline. If the asteroid approaches Earth at the opposite edge of its gravitational flux, it will be deflected away just the same by flux compatibility. One might wonder why asteroids do not approach Earth from the poles where the injection angle is much wider. The answer is: asteroids come from tails of comets or debris from planetary collision with planetoids or collision among wayward planetoids that lie in the Sun's dark halo along SEP. They are attracted towards massive planets by gravity. Some get suspended in neutral regions (*i.e.*, dominated by the Sun's gravitational flux). How can a planetoid go wayward? When a comet passes nearby it pulls the planetoids away from their orbits around the Sun, by flux-low pressure complementary.

In the case of Jupiter which has powerful gravitational flux, while it has also a natural shield against the massive asteroids or wayward planetoids, its powerful gravity overcomes their momentum to escape suction along the spiral streamlines and so they fall into the planet. We have seen this spectacle in the 1990s when large asteroids coming from the tail of a dying comet collided with it and caused powerful earthquakes detected on Earth.

We already know what happens to cosmic dust; they get entangled with cosmological vortices and collect at their cores.

Extension

Efficiency in a control system is improved by varying controls and this is where we need other families of spiral curves from the family of curves by shifting origins in a systematic manner to compose a simplicial trajectory corresponding to piecewise constant controls. In general a convex polygon in the same n-space will do for the control region. It can be dials on a discular control region switched from one position to another. In our case we took a parallelogram U with no side parallel to a coordinate axis.

We determine the directions of y orthogonal to sides of U and the values of t at which u(t) switches from one vertex to another. These are at corresponding intersections of curves $\eta(t)$, $\infty > t > 0$, with lines l_1, l_2 through the origin perpendicular to the sides of parallelogram U (note that we have shifted origins relative to the vertices). The η-curves are counterclockwise arcs of unit circle and the intersections occur at four sets of periodic values of t of period 2π. One of these families is a trajectory corresponding to constant controls determined by the origin; they represent streamlines. They are visible in the galaxies with visible matter, specifically, minor vortices as tracer. A streamline is the most efficient path of a particle (uncontrolled) to reach the origin.

To find the suspected optimal control we apply the integrated Pontrjagin maximum principle and the above Young's theorem [119] to the solution of (2):

$$y = ce^{-\lambda t}\zeta., \ \zeta = ae^{j(t+\lambda)}, \qquad\qquad (9)$$

where c and a are constants of integration and a > 0. We determine the directions of y orthogonal to sides of U and the values of t at which u(t) switches from one vertex to another. These are at corresponding intersections of curves $\zeta(t)$, > t > 0, with lines l_1, l_2 through the origin, perpendicular to sides of parallelogram U. The ζ -curves are counterclockwise arcs of unit circle and the intersections occur at four sets of periodic values of t of period 2π.

Without loss of generality we suppose l_1, l_2 to lie in the first and third quadrants, respectively, and denote by α_k (k = 1, ..., 4) the angular sectors on which l_1, l_2 divide the plane and by u_k (k = 1, ..., 4), the vertices of U. The indices indicate orientation: as we proceed counterclockwise around the origin starting from the real axis we pass from α_1 to α_2 by crossing l_1, then from α_2 to α_3 by crossing l_2, etc., (figure in [43]). For each k, the vertex u_k of U is in sector α_k and when l_k is in sector α_k the control u(t) along suspected optimal trajectory takes the value u_k. We term sojourn time at u_k the time that elapses between consecutive switches for which u(t) = u_k, *i.e.*, $\zeta(t)$ is in angular sector α_k, and the sojourn time is its measure which does not depend on the time and place at which the previous switches occur.

To avoid confusion due to the shifting of coordinates we now refer to the curves in (9) as the ξ - and φ –curves, respectively. To determine the trajectories, we denote by x_k (k = 1, ..., 4) vertices $-A^{-1}u_k$ of the parallelogram derived from U by the affine transformation u → x given by Ax + u = 0 or x = $-A^{-1}$ for which u_k satisfies

$$d/dt(x - x_k) = A(x - x_k), \qquad\qquad (10)$$

obtained by translating the origin to point x_k from corresponding φ-curve solution of dx/dt = Ax:

$$x = ae^{\lambda t}\xi(t). \qquad\qquad (11)$$

Here, *a* and λ are constants of integration and *a* > 0 except for the trivial solution $\varphi = 0$. The origin is reached for t = 0. The ξ - and φ -curves coincide and the latter is derived by multiplying by the factor $ae^{\lambda t}$ which deforms them towards the origin.

What remains is fitting together suitable translations of the φ-curves for which the origin has been shifted, in turn, to the appropriate vertex x_k to form the suspected optimal trajectory which, by the above Young's theorem is uniquely determined by the set U, is necessarily optimal. The shifting occurs at switching times at the intersections of the arrows and l_1 and l_2 (graphics in [41]). Also, by the Pontrjagin Maximum Principle, the terminal segment hits the target, boundary of a half-sphere with center at origin, orthogonally. The sequence of convex polygons, edges approaching zero length, as control sets generate the sequence of optimal trajectories converging weakly to the optimal generalized curve (absolutely continuous with set-valued derivative). Controls can be dials on control panel.

As matter accumulates at the core the spin increases due to momentum conservation and, with dark viscosity, extends its impact on wider region. The particle outside the core takes increasing centrifugal force which, combined with gravitational flux pressure, yields a resultant that serves as control with reverse effect: delays the transfer of particle to the core until the balance is attained between the centrifugal force and the gravitational pressure. Stability ensues with radial oscillation yielding elliptical orbit. If the centrifugal force critically exceeds the gravitational pressure the particle takes a parabolic orbit and escapes.

This physical solution can be mathematically modeled by a control system where the sequence of control sets U consists of convex polygons having arbitrarily fine edges and the objective function is suitably modified to generate the sequence of piece-wise constant controls u(t) and the corresponding trajectories weakly converging to a generalized elliptical trajectory (see figures). The conjugate curves representing the streamlines from which the sequence of optimal controlled trajectories is formed with controls induced by the resultant of gravitational flux pressure and centrifugal force at any point on the plane. However, the computational method (e.g., symbolic computation) applied to the solution of the pre-problem offers more efficient solution, the same method used in space exploration. Nevertheless, both physical and computational mathematics apply here only because the pre-problem has been solved and the solution is anchored on qualitative mathematics. Here we see the precise complementary relationship between computation and its complement, qualitative mathematics and, in terms of methodology, conventional and dynamic modeling. Qualitative modeling and its mathematical component, qualitative mathematics sets the physical basis for the problem, that gravity is dynamics of vortex flux of superstrings. Then computation finds the trajectory and locates the body along it.

Both the solutions of equation (2) and the adjoint equation (5) are parametric equations of two families of trajectories, one family spiraling counterclockwise towards and winding around the target circle with center at the origin and the other spiraling clockwise towards and winding around the same target circle. Each trajectory in either case corresponds to constant control, fixed parameter. Thus, control is applied by varying the value of the parameter, in effect, the derivative. The solution we got is not the optimal solution but an improvement over one with constant control, i.e., along a fixed spiral. A much improved trajectory is one with rapid switching between u1 and u2 that generates a simplicial curve analogous to the zigzag except that instead of mixture of line segments we have here mixture of spiral segments. The limit of a sequence of such simplicial curves, as the switching becomes more rapid indefinitely, is an infinitesimal simplicial curve, a generalized curve, along spiral trajectory. That limit is an optimal trajectory.

Any body B among the rest of the n bodies can be located by fractal transformation using the collected mass of the core vortex as the fractal generator. Body B takes an elliptical orbit when the gravitational pressure at B balances the centrifugal force on B combined with usual radial oscillation at the point of balance.

The main and secondary streamlines on the corresponding spins can be mapped. The main spin generally retains residual secondary spin (e.g., the Earth's oscillatory wobble known as summer-winter solstices) or takes minor spiral spin. The Earth's main spin is the 24-hour rotation. Around the polar axis at the opposite poles are conical spiral streamlines whose cross sections normal to the axis form concentric circles covering the plane with center at the axis of major spin (only Mercury is in one of those polar streamlines which accounts for its perihelion shift). All such spirals are dark streamlines towards the event horizon of the vortex eye. As the accumulated mass at the core cools off pressure at the eye reverses and exerts sucking action at the poles resulting in flattened polar caps.

Concluding Remarks

This solution is direct application of qualitative modeling: to solve a problem one devises a theory that provides the solution. Neither Newton's laws of motion nor Einstein's field equations can solve this problem. Since they are not qualitative but mathematical models they can only describe the behavior of a physical system but cannot explain them. Therefore, the equations have no predictive capability on the n bodies. Relativity has the basic ingredients of a theory but there are problems with the postulates.

GUT facilitates search for verification of or counterexamples to physical theories. For example, asteroids between the orbits of Mars and Jupiter and the larger ones around the corridor between the orbits of Neptune and Pluto [92, 108] are verification of its prediction that bodies in the Cosmos that have lost gravitational flux coherence (e.g., debris) lose gravitational interaction between them. Since asteroids have mass they are counterexamples to Newton's law of gravitation.

We note at this point that the gravitational n-body problem has different solutions when the n bodies are not collected mass at cosmological vortices (e.g., asteroids and comets). Let us consider an example of the extended solution of the n-body problem.

Suppose the problem is to land a rocket on the Moon. In this case the entire trajectory is mixture of segments of trajectories from the Earth to the Moon as follows: (a) trajectory from launch site to a point of tangency P with an orbit around Earth; (b) then the space vehicle goes around the Earth in this orbit a few times to prepare for "break-off" at a certain point Q in this orbit so that we have now a mixture of paths from launch site to point Q. (c) Then the space vehicle fires its rocket at point Q for the trajectory towards a point of tangency S at a lunar orbit; the space vehicle may go around this lunar orbit to survey the terrain at landing site. (d) Then at some point T on this lunar orbit the space vehicle launches the Lunar Lander towards, stops and hovers at some point Z, say, six feet above target and drops it on target (in this case the orthogonality condition is satisfied).

GUT TECHNOLOGY

A GUT technology is based on GUT and runs on the clean, safe, free and inexhaustible dark matter abundant everywhere in the entire Cosmos. Like the steam engine that preceded the science of thermodynamics by a century, there have been many operational GUT technologies long before the development of GUT that started in 1997. GUT technologies have a common thread: utilization of dark matter to generate kinetic energy. Perhaps, the oldest GUT technology aside from the battery is the electric power plant or, at smaller scale, the electric generator invented three hundred years ago based on the utilization of the magnetic flux, a vortex flux of superstrings around a magnet and a natural engine in dark matter.

A more recent GUT technology is the magnetic train powered by the magnetic flux of a magnet fixed on the rail and another underneath the train so that their equatorial fluxes are in opposite directions and, by flux compatibility, push each other and the train forward. Both magnets may be electromagnets taking power from the power line and the one on the rail turned on automatically by an approaching train so that power consumption is minimal.

There are many advantages of the magnetic train. Moving parts and grinding are minimal and so is wear and tear. With less friction it is efficient and speed is limited only by safety considerations. Most electrical power comes from the vortex fluxes of superstrings around magnets including atoms which are free, clean and abundant.

The electric motor is another GUT technology. In fact, there are, many operational GUT technology but they are still conventional in the sense that they were devised without benefit of theory. Their inventors relied on knowledge of properties of materials. However, with GUT, more powerful technology is on the horizon. Some of them are at the conceptual and prototyping phase. We shall refer to machine that runs on magnetic flux electromagnetic machine. One electromagnetic engine that is quite practical to make can run a car, bus or truck by utilizing vortex flux of superstrings in another way. Now, electric power generation of any megawatts of power is possible using the same principle limited only by space and safety considerations. This kind of technology converts dark matter to electricity directly.

Magnetic levitation occurs because the vortex flux of the magnet has an eye which creates low pressure in the magnetic flux of the Earth, which is also a vortex flux of superstrings, and so the magnet tends towards low pressure, the thinner layer away from Earth's surface and towards the gravitational flux rim far beyond the orbit of the Moon. The natural law behind magnetic levitation, flux-low-pressure complementarity, can be applied to revolutionize space travel including air travel. All it needs is put a guidance mechanism on a vehicle powered by it, say, a gyroscope, and control of the magnetic flux by scattering and containment and one can have a vehicle that can travel at great speed within the Earth's gravitational flux. Beyond the Earth's gravitational flux is the Sun's, beyond it, the Milky Way's, etc.

One advantage of dark matter as fuel aside from being free, inexhaustible and abundant everywhere in the Cosmos is it does not tamper with the ecology and fits the requirement of sustainable development [6]. How can we apply it to practical affairs?

In the fifties the International Monetary Fund pushed a development scheme for Asia called Import Substitution. It failed. Why? The scheme aimed at providing self-sufficiency in technology by developing it locally so that importation of technology is minimized. This is easier said than done because in an underdeveloped country both the government and the people do not have the resources to fund development of technology nor do they have the capacity to buy its products. In other words, development cannot be undertaken from within. The developed countries of today took centuries to develop because even their colonies were underdeveloped and could not contribute significantly except their cheap raw materials which were inadequate to pay for the cost of development.

Consequently, that scheme was a total failure; worse, the countries that adopted it not only failed to develop but were also buried in the quagmire of foreign indebtedness under the auspices of the IMF-World Bank and they are still reeling from it now. The only countries and territories that succeeded were those that rejected it and embarked on their own development even if based mainly on common sense without formal strategy. They are now referred to as the newly industrializing countries (NIC) and territories of South Korea, Singapore, Taiwan and Hong Kong. Of course, that Club has now expanded with China joining in.

The remedy for the failed Import Substitution scheme is Strategic Positioning. Its lynchpin is creation of a couple of technologies that can capture global monopoly of the market and have a substantial share of the enormous global resources to spur local development of an underdeveloped country, specially, the needed infrastructure and provide the resources for the purchase of hundreds of other technologies for production to support the necessary infrastructure development and provide for domestic consumption and social services. There is a track record for this strategy in the success of the NIC. We elevate their experience into a formal strategy.

The lynchpin technology for Strategic Positioning is no longer available in high technology which is crowded and now monopolized by the developed countries and the NIC. This is where we turn to GUT that has brought us to the threshold of a new technological Epoch. The strategy is not exclusive to any country because the technological possibilities are wide open. We are just at the threshold of GUT technology at this point and any country can find a niche there.

CHAPTER 5

Scientific Natural Philosophy

Abstract: Having laid down the mathematical and physical foundations of thought through the theory of intelligence we have presented scientific natural philosophy and its mathematics briefly but comprehensively. Since knowledge is dependent on methodology, we have given methodology and epistemology the same importance although the latter is the core of philosophy by tradition. Being materialist, scientific natural philosophy relies on rational thought. Therefore, we undertook critique-rectification of mathematics and its foundations and the methodology of natural science to minimize the inherent uncertainty of thought.

Unlike previous philosophies scientific natural philosophy is far from being speculative based on the most advanced findings of natural science, particularly, the discovery of the basic constituent of matter, superstring. Its discovery was the key to our knowledge of the other inhabitants of our universe from micro/quantum through macro/galactic scales, and from dark through visible matter. For the first time, we know what a black hole and a big bang are. We are now able to explain wave-particle duality, metal fatigue, elasticity and superconductivity. We know the evolution of our universe from its birth at Cosmic Burst through its destiny as a cluster of black holes back in dark matter and the place of our universe in the timeless and boundless Universe – one of the bubbles in it. Moreover, natural laws are transitory which is quite contrary to traditional thinking. There was no natural law in this part of the Universe before the Big Bang, and all natural laws will vanish as our universe reaches its destiny.

INTRODUCTION

Philosophy strives to provide the deepest and broadest knowledge of a given discipline. This is the reason one who achieves such depth and breadth of knowledge and makes outstanding contribution to a discipline is awarded the Doctor of Philosophy degree. The preceding chapters have provided not only such depth and breadth of knowledge of nature and its language – mathematics – but also its applications both theoretical and practical. This leaves our main task – elaboration of scientific natural philosophy – lighter. We simply need to summarize what has been done earlier, identify new knowledge and highlight the resolution of important issues in science, mathematics and philosophy. We have utilized all the tools at our disposal for this purpose. Now we highlight the new knowledge we have gained and identify the jump-off points from the old mainly in terms of the rectification and advances we have delivered.

The core of philosophy is epistemology or theory of knowledge that represents what happens in thought. However, since knowledge is dependent on methodology we treat it at the same level of importance as the complementary methodologies of qualitative and quantitative modeling. The principal tool and component of qualitative modeling is, naturally, qualitative mathematics, that of quantitative mathematics computation and measurement. We now formally identify qualitative mathematics as the representation of rational thought subject to the requirements of a mathematical space identified earlier and forge the most advanced methodology of natural science – qualitative-quantitative modeling. Just as intuition is the driving force of idealist philosophy, rational thought is the dynamo of materialist philosophy.

We distinguish thought or intelligence from mind which includes feelings and emotions centered in the heart. Both individual knowledge and intelligence are centered in the CIR. We know that while quantitative modeling only describes the appearances of nature, qualitative modeling yields genuine knowledge beyond appearances by providing explanation of how nature works. In this sense, methodology is more important than epistemology, the reason we provided full discussion of the mathematics of GUT, especially, the advances we have made in ridding mathematics of its ambiguity that harbors contradictions.

ASSESSMENT OF THE MATHEMATICS OF GUT AND ITS FOUNDATIONS

We summarize the major rectification and clarification made on the mathematics of GUT. All of it belongs to the methodology of natural science – qualitative-quantitative modeling.

(1)   We re-asserted the clarification made by David Hilbert regarding the ambiguity of individual thought stemming from its inaccessibility to others and the fact that some of its concepts cannot be captured by its representation in the real world that makes it necessary to build a mathematical space on objects in

the real world well defined by consistent set of premises or axioms. Thought is capable of conceptualizing events that cannot happen or geometrical figures that cannot exist in the real world. For example, one can think of a snake as a physical object that swallows itself from the tail up completely but it is impossible to figure out its mathematics. We can also conceptualize travel from Earth to Andromeda in an instant which is impossible in the real world. In other words, there is complete freedom of thought but the real world is subject to the laws of nature. Another example is sensation which cannot be accurately and objectively represented in the real world. We can only establish some correspondence between two individuals' sensations.

(2)   By and large, this clarification has not been grasped by mathematicians as shown by the great controversy around the mathematical objects 1 and 0.99… for over a decade. Most mathematicians accept the equation $1 = 0.99…$ which is akin to the equation apple = orange that says distinct objects are the same. Only the same objects can be equated to each other so that the relation "=" means "the same as". In other words, the symbol "=" can only be used reflexively. How about the statement "$2 + 3 = 5$" where the left and right sides of the equality are different objects? Is it correct? Yes, since the left side is a binary operation or mapping, its result or image being the object on the right side of the equation.

(3)   We have also identified the sources of ambiguity that need to be avoided in the construction of a mathematical space. They include infinity, vacuous or ill-defined concept, self reference and large and small numbers, to mention a few. Moreover, any categorical statement about ambiguous concept, e.g., the use of the universal or existential quantifier on infinite set, is ambiguous. In this regard, since the axiom of choice and its variants, the completeness axiom and Dedekind cut are categorical statements about the infinite set of real numbers all theorems that depend on them in their proofs collapse, e.g., the Heine-Borel and compactness theorems in the real line. The Banach-Tarski paradox blamed on the axiom of choice [75] is really due to the ambiguity of infinite set; the axiom of choice is only incidental. This clarification puts some restrictions on the construction of a mathematical space. For example, to avoid ambiguity building a mathematical space must start with finite objects and, when necessary, expand it by admitting only contained ambiguity, *i.e.*, approximable by certainty. For example, a nonterminating decimal is approximated by its initial segment with desired margin of error. A nonterminating decimal is inherently ambiguous even if there is an algorithm for computing every digit because not all of its digits can be computed or identified. However, it is contained since it is approximable by a decimal segment at any desired margin of error.

(4)   Since nature optimizes and works in accordance with the principles of non-redundancy and non-extravagance and given that mathematics is a language of natural science and must reflect its nature and meet its requirements, there is no need to develop set theory apart from existing mathematical spaces. Moreover, a set is interesting only when it has structure provided by some axioms in which case it is a mathematical space. The tradition in teaching is to introduce a mathematics course with discussion of set theory which vanishes from the course as it progresses; it is an unnecessary ritual.

(5)   We have replaced the ambiguous infinitesimal of calculus with the well defined dark number d*, infinity with the unbounded number u*. The dark number d* joins all the decimals together into a continuum that, together with the unbounded number u*, comprises the new real number system **R*** which is a continuum. Our construction of **R*** retains all the valid, useful and interesting properties and concepts of the real number system **R**. The only exceptions are the Archimedean property and its nondenumerability which do not hold in **R***. However, the decimals form a countably infinite subset of **R*** and since the construction of the Banach Tarski paradox depends on the Archimedean property it does not arise in **R***. However, the decimals retain both the Archimedean and Hausdorff properties. Therefore, they retain all the correct properties of the real numbers. The problems that could not be resolved in **R**, namely, Fermat's and Goldbach's conjectures have been resolved in **R***. We have also pointed out the error in Cantor's diagonal method in proving the nondenumerability of **R**. As of this moment there is no non-denumerable set, only countable infinity exists. However, we predict that no non-denumerable set can be constructed because physical objects in our universe are finite. Cantor's diagonal method generated only countable set of off–diagonal elements. What they are is not even identified.

(6)   We have well-defined the nonterminating decimal for the first time as the g-limit of its generating sequence. There are only two kinds of decimals – terminating and non-terminating. The concept *irrational* is ambiguous because there is no way to verify if a nonterminating decimal has nonperiodic decimal expansion.

(7)   Since a mathematical space is well defined only by its axioms, it is fallacious to use external rules of inference on it like formal logic; the rules of inference must be well-defined by its axioms, *i.e.*, any conclusion drawn from it must follow from them.

(8)   We have explained that the axioms of a mathematical space do not well define its extension; therefore, the latter requires a new set of axioms. However, an extension must be consistent with its restriction to the base subspace, *i.e.*, the original space that was extended.

(9)   We showed the inconsistency of the field axioms of the real number system and constructed the contradiction-free new real number system on the additive and multiplicative identities 0 and 1 based on three simple axioms.

(10) We identified the contradiction of the vacuous concept i of the complex number system, replaced it by the operator **j** on plane vectors and constructed the contradiction-free Complex Vector Plane. Moreover, the latter mathematical space has given rise to a new binary operation on vectors that yields a vector, namely, vector multiplication distinct from the vector and scalar products.

(11) Some local mathematical principles or properties do not necessarily hold globally. For example, the solution of a differential equation is valid only at a neighborhood of a point.

(12) Above all, we have introduced qualitative mathematics and modeling and deepened scientific knowledge from description of the appearances of nature to explanation of how it works based on natural laws and systematically organized such knowledge into a physical theory – GUT – built on them. This theory is tailor fit for theoretical and practical applications in the various fields of natural science. In general, theoretical application requires additional discovery of appropriate natural laws in the fields concerned. Practical application of theory includes devising technology and GUT opens up a new technological epoch that we call GUT technology. A GUT technology runs on the clean, free and inexhaustible dark matter abundant everywhere in the Cosmos. There are, of course, technologies belonging to the new epoch such as the magnetic train and electric power plant but they are just the tip of the iceberg, so to speak, and are still considered conventional since they were devised based solely on present information about the behavior and properties of materials not on GUT. The situation is similar to the invention of the steam engine that preceded thermodynamics.

GUT: PATTERNS, SIGNIFICANCE AND VERIFICATION

In the new science articulated by GUT we note the following:

(1)   Symmetry and duality in nature. The latter may be considered reverse symmetry. For example, quantum gravity is the dual of macro gravity where the atom is the dual of the galaxy, matter-anti-matter interaction that of supernova and the orbital electron that of a star.

(2)   Some properties of matter are individual, others collective. For example, speed and momentum are specific to individual physical systems but pressure, temperature and humidity are collective properties. They are not obtained from their individual properties. In the same way, the capability of a social organization is not the sum of the capabilities of individual members.

(3)   Some principles that apply to local behavior are reversed in the global setting. For example, the second law of thermodynamics applies only to local closed systems like enclosed gases where entropy or disorder tends to rise. At the same time, our universe as a closed system has evolved from the chaos of the Big Bang and Cosmic Burst through the present higher order including the biological order that we now enjoy.

(4)   We have rectified a contradiction in conventional physics – the existence of massless matter like the superstring and photon with clear definition of energy as motion of matter; mass is a measure of the amount of matter in any physical system.

(5)   Traditionally, theory is viewed as mere formality and derided by the familiar remark, "that is only a theory". Now we have placed physical theory, specially, GUT in the right pedestal: at the center of natural science that glues and unifies it. There is no knowledge without theory only hodgepodge of information.

(6)   In this new science the primary verification of a physical theory is its being able to explain natural phenomena. In current science it is correctness of its prediction. There are instances where correct

prediction is based on wrong premise. For instance, Einstein predicted correctly the inward deflection of a ray of light that passes near the Sun. His explanation was that mass (in this case the Sun) creates curvature around it. Both that curvature and how the Sun creates it are not known. Now we know that the curvature of space around the Sun is really its gravitational flux. Its relationship with the Sun is reversed: the Sun did not create its gravitational flux; rather, it was this gravitational flux that collected the mass around its eye we call the Sun by flux-low-pressure complementarity.

(7) Unless a physical theory is able to explain natural phenomena its prediction is worthless. Once primary verification is attained, technology that works based on physical theory is weightier than prediction and so we consider technological application the second level of verification, correct prediction the third. Moreover, devising a piece of technology involves some prediction.

(8) As we already noted while conventional cosmology presents an unstable universe subject to the anthropic principle [74] GUT's cosmology is stable and our universe as generalized nested fractal sequences of cosmological vortices with our universe as the common first term is stable and every cosmological vortex in it is stable and has specific destiny.

We mention in passing that any explosion is a point of discontinuity and deletes all information about the history of the physical system that exploded. Two such explosions occurred in the Cosmos: the Big Bang and the Cosmic Burst. Therefore, the claim that background radiation verified by the COBE project [106] came from the Big Bang is not valid for whatever came from it was demolished by the Cosmic Burst.

Ultimately, it is nature, not the scientist that confirms the correctness or validity of a physical theory when all three levels of verification are achieved; clearly, we have raised the standards for verification of theory.

The energy conservation equivalence law is next in importance to energy conservation. It pins down the structure of important physical systems, e.g., that of the primum and its sinusoidal profile which also applies to the profile of any wave including water and cosmic waves. Now we can also use it to explain why the photon is a rapid oscillation. Since its toroidal flux has enormous speed of 7×10^{22} cm/sec and its forward speed (forward component) must equal the speed of the carrier basic cosmic wave (which is the speed of light of $c = 10^{10}$ cm/sec) to be stable (otherwise, it will leave the carrier and disintegrate since it has broken away from its superstring loop) it is only possible if its oscillation is suitably rapid, *i.e.*, its arc length is suitably fine. This also explains, at the same time, why the photon has no rest mass.

Another philosophical point related to oscillation is that motion breaks symmetry and symmetry precludes motion. Oscillation is the optimal balance between them. Moreover, since micro and dark scale vibrations are universal (every physical system vibrates due to the impact of cosmic waves coming from all directions) and the graph of vibration plotted against time is a wave (sinusoidal) then a physical system in motion resonates with this vibration to overcome resistance. The non-redundancy and non-extravagance principles also clarify a theoretical point. In 2004 there was much excitement when the third quark was discovered in the nucleus outside the proton for it was thought to be the gluon that supposedly binds the protons in it. This author was then the science editor and a columnist for the Manila Times. He got e-mails from readers asking for comment. The negative quark joins the two positive quarks of the proton; so does the third quark which joins two positive quarks one from each of two protons. By these principles of energy conservation equivalence the third quark is a negative quark since it does the same function as the latter. Therefore, the gluon is really the negative quark discovered at FermiLab in Battavia, Illinois several decades ago. At that time the search for the irreducible elementary particles had been completed, thanks to particle physicists. They are the electron, the +quark and the –quark that comprise light isotope of every atom. A heavy isotope has one more component – the neutrino – consisting of two prima of opposite but numerically equal charge; what those prima are still needs to be determined by particle physicists.

Since qualitative mathematics creates physical theory there is a compelling reason to raise standards. We have seen that mathematics, particularly, qualitative mathematics, is the glue that binds physical theory together; when it fails it is bound to be disastrous. In fact, catastrophic failure in the real world is ultimately linked to theoretical error or inadequacy. One example was the disastrous final flight of the Columbia Space Shuttle of February 1, 2003 [40]. We also know of supersonic jets that mysteriously explode in mid-air. Both phenomena have never been explained by conventional science and the reason was lack of adequate theory to provide the explanation due to failure of qualitative mathematics. Now, with GUT, we know that the former disaster was due to the micro component of turbulence that generated and propagated seismic waves that softened the metallic attachments of the Shuttle. There are other cases of failure not as dramatic but have drained great amount of resources for research without yielding the expected results. A case in point is the smashing of the nucleus of the atom which has been going on for over half a century. The problem here stems from unclarity of objective. If it aims to discover the irreducible elementary particle, it has been achieved with astounding success by 1990. All they have found besides the basic prima and neutrino are man-made particles in the laboratory that vanish in split second which means that nature does not need them, by the non-redundancy and non-extravagance principles. Therefore, none of them is even a candidate for basic constituent of matter which must necessarily be indestructible. In fact, GUT says that the basic constituent cannot be found in the nucleus because everything there appears destructible.

Another case of theoretical inadequacy is fusion research which has similarly been going on for over half a century for the purpose of generating energy like the nuclear reactor does. The premise is that if two atoms of hydrogen can be fused together into helium it will release energy equal to the difference between their atomic masses. The great release of energy in thermonuclear explosion is attributed to such fusion. But it is almost impossible or prohibitively costly to fuse the nuclei of two atoms because of the enormous repulsion between their protons. It is reverse alchemy which has been done only at great cost. Moreover, there is no law of nature that justifies fusion of two nuclei. Again, the failure of this project is due to incorrect or lack of theory. The great release of energy in the explosion of the hydrogen bomb is due to the agitation of the superstrings in the nuclei of the hydrogen fuel by the explosion of the trigger atom bomb that converts them to visible matter and energy.

We have focused on the three pillars of GUT, namely, quantum and macro gravity and thermodynamics. The crucial factor for the development of GUT is the introduction of qualitative modeling for the discovery of the appropriate laws of nature that provide the axioms of these three pillars as subtheories of GUT. The most fundamental natural law is energy conservation that manifests itself everywhere in nature in accordance with the energy conservation equivalence law. The latter law is still partial and keeps increasing as our experience increases. So far, the energy conservation equivalence law enumerates about 25 manifestations of energy conservation. Optimization as a manifestation of energy conservation extends to the ecology in terms of optimization of waste. For example, tissues left on the bones of the lion's prey are cleaned up by the ants. There are, in fact, many instances of optimization in nature.

For the first time quantum gravity has provided answers to the fundamental questions raised earlier such as the structure of the superstring, electron and atom. We have also explained many physical concepts and physical interactions that constitute new knowledge such as the primum and its charge, matter-anti-matter interaction, photon, wave-particle duality, cosmic wave, formation of heavy isotope, elasticity and metal fatigue. Moreover, we have corrected misinterpretation of thermonuclear explosion. It is due not to the fusion of hydrogen atoms but to agitation of the accumulated superstrings in the hydrogen nuclei by the explosion of the trigger atom bomb. This accounts for the lack of breakthrough in research and development on thermonuclear reaction after more than 50 years.

Among the important contributions of thermodynamics is full explanation of the disastrous final flight of the Columbia Space Shuttle. However, it required the discovery of new laws of nature to do so. Since thermodynamics in the broad sense includes generation of kinetic energy, *i.e.*, conversion of dark or latent energy to visible or kinetic energy its further contributions include practical applications, specially, technology generation.

Macro gravity has provided full explanation of physical systems in the Cosmos from the super…super galaxy, our universe, through the galaxies, stars, planets and cosmic dust. Most of all, it has provided the cosmology of our universe from its birth at Cosmic Burst through its evolution to the present. Since the destiny of every cosmological vortex is a black hole in its eye [25, 37, 42, 100, 104] the destiny of our universe is a cluster of black holes back in dark matter.

DIMENSION

In a vector space dimension is easy to define: the cardinality of its basis or, alternatively, the maximum number of linearly independent vectors in the space. Mathematical dimension can be infinite. Thus, the dimension of the Hilbert cube is countably infinite.

In physics this concept is quite vague and sometimes viewed with mysticism. For instance, a block hole is supposed to suck matter from our universe and brings it to another dimension. The space of relativity is generally considered four-dimensional, three space dimensions and time. However, the Lorentz transformation reduces it to three for it establishes dependence between time and a space coordinate. How do we well define this notion in physics? We first make some clarification. We qualify that a physical concept exists in the real world, e.g., some object including symbol or measurable entity like temperature, pressure and humidity, which are not man-made concepts of thought. How about time? Is it a physical concept? Obviously, not since we do not find it in the real world. It is a man-made concept of thought that expresses relationship between events. So is the concept *distance* that expresses a relationship between physical objects or events. However, as mathematical concepts they describe natural phenomena. In particular, they can be used for devising a coordinate system. Thus, dimension depends on the subject of investigation. In meteorology, for instance, we can have the three space dimensions plus time, temperature, pressure, visibility and humidity for a total of 8 dimensions.

Since every physical system is bombarded by cosmic waves from all directions which resonate with the superstring, the latter's vibration has large dimension in the dark region of matter. At the micro (atomic scale) and macro scales, however, the dimension of vibration is limited by the internal-external dichotomy law [42]. For example, the nucleus of the atom vibrates normal to its equatorial plane. Therefore, its motion has four dimensions, *i.e.*, one more than the three dimensions of ordinary space. Cosmic waves do not resonate with physical systems beyond the micro scale and, therefore, has only indirect measurable impact on natural phenomena.

THEORETICAL APPLICATIONS OF GUT

We first go through the theoretical and practical applications of GUT for informed philosophical analysis and summation. Theoretical applications include opening up new fields of science.

Theoretical Application: The Unified Theory of Evolution

We have updated Darwin's theory of evolution based on GUT. We note that biology has the most advanced application of what we call naïve qualitative modeling, namely, classification of evolutionary paths of the various species to identify certain patterns in their evolution; qualitative because it employs minimal computation and measurement but naïve because it is descriptive. Recently, biology has begun to employ quantitative mathematics and modeling but not as advanced as mathematical physics does. Now, for the first time, we have fully employed qualitative modeling and advanced our knowledge of the subject. Among the new knowledge in addition to the biological laws are the following:

(1)   The brain and the gene use the same medium for their functions: brain waves. The main function of the brain is control of body functions including the immune system and its secondary function is thought centered in the cortex which is possessed only by higher species including but not exclusive to cats and primates. (The author had a graphic but unpleasant experience of how his pet cat avenged its being thrown out of the bedroom and waited for him to come out of the door to know what it had done and ran away) The main function of the gene is production of the tissues of an organism in the cellular membrane. The gene is also the agent of heredity, its secondary function.

(2)   The principal factor in the evolution of a biological species is the gene; in an organism a gene may be modified, altered or acquired at any point in its evolutionary path and we indicated how this can occur including self-induced genetic alteration. This is the basis of possible splitting of species. For example, the initially clumsy snake started with a pair of legs at its tail. Then the legs were discarded in favor of agility and speed of movement of the wiggly snake. Then there was a split, one species going for size and strong muscles of the anaconda and python. Then a split occurred again that gave rise to species of small but venomous snakes, now the dominant species. These advances are based on genetic alteration, modification and enhancement for species to gain advantages in coping with nature and competition for dominance over and control of other species. (One implication is that a genetic disease – its symptoms encoded in the gene – can be acquired but no gene can be removed because it is in the nucleus of every cell)

(3)   Evolution is much more than natural selection or survival of the fittest; through genetic alteration, modification and enhancement, a species not only strives to cope with the environment but also to

control it and gain dominance over other species. It is not only the human species that has acquired this skill. For example, some ant species domesticate others to gather food for them. Even some plants have gained dominance over some animal species, e.g., the mice eating pitcher plant of the Philippines. Its pitcher-shaped trap contains enzyme that decomposes the prey which is then absorbed by the plant. Some plants are insectivorous. A known insectivorous plant has a pair of sticky pods that closes when an insect perches on it. With enzyme in the pod the insect decomposes and gets absorbed by it.

Other Theoretical Applications

We applied GUT to develop the theories of chaos and turbulence, global geology and oceanography, biology and intelligence based on natural laws. The theory of intelligence required the discovery of new laws of nature to develop. So did the explanation of the disastrous final flight of the Columbia Space Shuttle that required the discovery of some natural laws. Among the new knowledge gained in the development of chaos and turbulence are (a) the existence of the micro component of turbulence at its interface and (b) its generation of seismic waves, (c) understanding of how turbulence occurs and how it can be controlled, fully or partially and (d) the relationship between chaos and turbulence in the standard dynamics. With respect to the disastrous final flight of the Columbia Space Shuttle at least two natural laws were discovered and needed to provide the explanation of why and how it happened. They are: resonance and existence of threshold for generation of seismic waves. The softening impact of seismic waves on malleable material (e.g., the chassis and metallic attachment) was also verified by the gradual separation of the Shuttle into several pieces.

The development of the theory of intelligence led to the discovery of the laws of thought and learning principles that have some bearing on mathematics-science education. The gene has much to do with intelligence in the sense that the latter's physical components, e.g., quality of neural vibration including retention of vibration characteristics, are genetic. For example, photographic memory is due to neural vibration characteristics that are not significantly eroded with the passage of time. Non-physical component of intelligence, e.g., creativity, reasoning skill and right approach to research is the result of proper training and substantial experience and practice. They make the difference between achiever and non-achiever. For example, there are bright mathematics students (e.g., child prodigy) who do not become outstanding mathematicians. However, there are average mathematics students who become great mathematicians suggesting that unless one has mental disability, e.g autism, he has as much chance as anyone else in harnessing the full potential of the human brain. Even some mental disadvantages, e.g., autism, can be overcome by proper therapy. In other words, good neural physical quality, which is genetic, can only contribute to one's intellectual potential; the rest is taken care of by proper training, creativity, right values or attitude and initiative.

PRACTICAL APPLICATIONS

Our first physical application of the methodology of GUT, qualitative-quantitative modeling, was the solution of the gravitational n-body problem that catalyzed its development and discovered the initial 11 natural laws of GUT. We summarize some of its technological applications.

With GUT and its methodology of qualitative-quantitative modeling we elevated other fields of science to knowledge beyond description of appearances of nature. Moreover, GUT provides a broader perspective for traditional inventors who rely only on information about properties of materials. For example, the information that cutting across magnetic lines of force by a conductor produces electric current has been the basis of electrical power generation in the last 200 years but understanding this phenomenon through GUT has made it possible to invent more powerful technology. GUT's explanation of magnetic levitation provides the key to the development of an engine that can revolutionize travel within the gravitational flux of the Earth and even beyond. Engine of this type is called electromagnetic. Electromagnetic engines are powered by clean free vortex fluxes of superstrings that do not pollute the environment. They are natural engines in dark matter with useful impact on visible matter. The same electromagnetic engine can generate electric current to any megawatt of power limited only by space and safety considerations. This can resolve the growing scarcity of electric power that in some places has reached crisis proportion.

Knowing how the gene works and theoretical knowledge provided by GUT through its application to biology paved the way for devising technology for the treatment of genetic diseases like cancer, systemic lupos erythematosus, diabetes, muscular dystrophy and mental disorder, e.g., depression, without injury to normal cells and, therefore, without side effect. It will involve genetic modification or alteration to remove their harmful symptoms [22]. In the case of cancer genetic sterilization is more appropriate at its early phase to prevent it from spreading and render it harmless. Sterile cancer cells are harmless and replaced during the normal turnover of cells if not destroyed by the immune system. Cancer destroys the body by reproducing fast and creating lumps that push and build pressure on normal cells around them and destroying the tissues.

Scientific Epistemology

With the theory of intelligence we can now consolidate and shorten our discussion of epistemology.

Rational Thought

Materialist philosophy relies on rational thought for its pursuit, idealist philosophy on intuition. However, rational thought also needs rectification. Traditional mathematics uses symbolic or formal logic to draw out conclusions from a set of premises. This way of reasoning has been systematized into mathematical logic. There is a flaw, however, in its application to a mathematical space or physical theory, which is well defined by its axioms or natural laws alone, since it is external to it and has nothing to do with the axioms. For this reason, we have replaced it by qualitative mathematics. Qualitative mathematics not only frees mathematics from the ambiguity arising from rules of inference not based on the axioms but also from the restrictions of deductive reasoning based on logic external to it and, therefore, releases thought for a myriad of other activity including theoretical exploration, reflection and creativity and ingenuity.

Science uses causation as its tool for rational thought. It assumes causes for natural phenomena in the conventional sense, i.e., appearances of nature. There is a lot of ambiguity in this approach, especially, when our knowledge is still partial and the causes of certain phenomena are not generally known. Again, we cite the deflection of a ray of light that passes near the Sun observed during solar eclipse which has been incorrectly attributed to the unknown curvature of space supposedly caused by the mass of the Sun which is not true. Both curvature and how the Sun's mass causes it are quite ambiguous because traditional science has no tool for explaining natural phenomena. Now we know that Einstein's curvature of space around the Sun is its gravitational flux and its mass has nothing to do with it. In fact, it is the other way around: the Sun as a cosmological body is the collected mass around the eye of its gravitational flux. There are a number of natural phenomena that has been attributed to the wrong causes, e.g., the energy released in thermonuclear explosion and the stability of the nucleus of an atom with more than one proton despite the enormous repulsion between them. Therefore, we have replaced causal reasoning by qualitative mathematics which, as in mathematics, frees thought for more creative activity. Qualitative mathematics is tailor fit for building scientific knowledge as physical theory since the latter is a mathematical space where the axioms are laws of nature.

Creation of Knowledge

Why do we *create* knowledge of nature rather than discover it? Natural laws upon which we build knowledge as physical theory are not found in nature. Only appearances of nature including natural phenomena are accessible to us. We articulate the laws of nature based on our interpretation, synthesis and analysis of its appearances and recognition of patterns of behavior. Such articulation is a creation of thought. Then we build scientific knowledge as physical theory based on it. There are many concepts which are creations of thought in the sense that they are not found in the real world. *i.e.*, they do not have physical referents. Among them are *time, distance* and *color*. In the case of color it consists of cosmic waves with specific wavelength reflected or generated by or having passed through and been modified by the natural vibration of an object. Both mathematics and physical theory are creations of thought based on interpretation, synthesis and recognition and analysis of patterns of behavior observed in natural phenomena through rational thought.

The validity of a physical theory depends on its adequacy in explaining natural phenomena, usefulness in guiding invention that works and verification of its predictions. In this regard, GUT brings us to the threshold of a new technological epoch. Some technological inventions, e.g., the electric power plant and the magnetic train, were made before the development of GUT just as the steam engine was devised before the emergence of the science of thermodynamics. In fact, traditional inventions are not guided by theory but based on extensive information about properties and behavior of materials.

The basic ingredients of knowledge are sensations modeled in the CIR by appropriate neural network. Their characteristics are encoded as corresponding vibration characteristics of their model neural network. When a concept is in thought its neural network vibrates its characteristics; once out of thought it reverts back to residual vibration, its components eventually stored in appropriate sensation regions. Storing encoded neural network in the CIR involves suitable vibration of its neural network that sends its components to their respective sensation regions, by energy conservation and resonance.

The component network gets to the right sensation region by resonance, *i.e.*, the neural path has the same vibration characteristics and order of magnitude of frequency as that of the sensation component's neural model. The CIR recalls sensation components of a concept by suitable vibration of its nodal region that flushes them back to the CIR, by energy conservation, for re-composition and activation of the concept at its nodal region. The nodal region is suitably connected cluster of neurons to every component of a concept, *i.e.*, its neural network. In other words, recalling a concept involves recalling its component neural network in the CIR and recomposing it at its nodal region. Note that sensation occurs in the CIR; the sensation regions only store components of concepts and there is no sensation there. For example, one cannot recall a whistle without shape even if sound and shape are stored in different sensation regions.

When a concept is not recalled for a long time it may become part of the subconscious and may not be recalled again. However, psychologists have developed some technique for recalling it based on association with contiguous events. The effect is to vibrate their nodal regions that, in turn, vibrate the nodal region of the concept in the subconscious and recalls its components for re-composition at its nodal region in the CIR. Moreover, psychologists have developed techniques for blocking recollection of certain events or altering recollection of events.

When a concept's nodal region is suitably agitated by basic cosmic waves from the Cosmos, its components are recalled and recomposed there; then that concept is activated and recalled in thought. This happens in spontaneous recollection. The CIR has the capability to recall a concept at will by vibrating its nodal region suitably which also vibrate their components sensation regions and, by energy conservation, flushes its component neural networks back to the CIR for recomposition at its nodal region.

The intensity of sensation is proportional to the size of the contiguous parallel bundle of neural chains or neural bundle that carries its components to the CIR to register and recompose at the appropriate neural region. Then its neural model consists of parallel contiguous neural network. However, by energy conservation and the principles of non-redundancy and non-extravagance only one full neural network that models the sensation is retained the rest de-activated and ready for encoding with another sensation. Intense sensation induces physiological change. For instance pain sensation produces molecule in the neural membrane. This was verified recently at the Pavlov Institute for Physiology [76, 84].

Concepts are components of thought and suitably interconnected concepts through their nodal regions constitute an idea or a proposition. Every join of a neural network of a concept is a nodal region. Then a system of thought must have a central neural region to which the rest of the neural regions are connected by bundles so that the central neural region is connected to every nodal region of every component. The neural regions of related concepts, say, in a proposition, are connected in the same manner except for scale; there are more nodal regions involved but it has also a central neural region connected to each of the rest of the nodal regions and, therefore, component network by a neural bundle.

Note that the neural network is fractal like the roots or branches of a tree connected to the stem corresponding to the central neural region. One can extend this network or cluster of concepts to a discipline of knowledge with the corresponding central nodal region, the intersection of the neural bundles of neural chains connected to all the components. Any anomaly in neural interconnection can cause common learning disability (distinct from mental disorder) such as changing the order in writing dictated telephone numbers, continuing on different line when reading a written text or missing or repeating phrases in typing. This type of learning disability can be controlled (in fact, the author suffers from some learning disability; most people do).

In the US, learning disability is recognized as a problem so that students with this problem form clubs to provide remedy just like Alcoholic Anonymous for alcoholics. A student with this problem can take appropriate precaution to cope with it. For example, he must sit at a particular spot in the classroom to minimize its impact. Once controlled, however, the person with this problem becomes relatively smarter as if some undesirable mental load has vanished.

How are values and way of reasoning formed? (We consider way of reasoning value because it is formed in the same way as common value) Values determine one's response, judgment or decision when a certain situation is at hand or, in the case of logic, when the hypothesis or axiom is given. A value or way of reasoning is formed in accordance with Pavlov's theory of conditioning. Correct choices or conclusion or judgment are induced by pressure or formal training. Pressure comes from the parents first, then teachers, peers and experience, e.g., in school or at work. Physically, it involves establishing neural connection from the central nodal region of the given condition or hypothesis or axiom to the nodal region of the right choice, judgment, decision or conclusion. In the neural model of a mathematical space, the axioms are neurally connected to the known theorems or principles through their central nodal regions. In science, all components of knowledge are neutrally connected to the natural laws through their nodal regions.

In deductive reasoning the hypothesis is neurally connected to the conclusion. In inductive reasoning, however, a single given condition may be neurally connected to a number of possible conclusions through their central nodal regions. Inductive reasoning is only a part of qualitative mathematics in the complement of deductive reasoning.

Since scientific knowledge of a particular field is articulated by a physical theory which is a mathematical space whose axioms are laws of nature then our knowledge of nature is the totality of physical theories we have already built and subsumed under GUT so that in the neural network model of GUT its neural nodal region is the central nodal region to which the neural nodal regions of the various fields of science are connected in a pyramidal hierarchy with the central nodal region as apex. Again, its neural network is a (generalized) nested fractal with all components issuing from its neural central nodal region through their nodal regions. Then values are modeled by interconnected neural network through their neural region not necessarily forming a pyramidal hierarchy but a discipline of knowledge is.

A discipline of natural science is a subtheory of GUT, *i.e.*, it is well defined by a subset of the natural laws of GUT; a field of a discipline of GUT, its subtheory, is well defined by a subset of the natural laws that well define that field. The neural model of every field includes all the neural models of the known consequences of its natural laws. Here we can see the nested fractal structure of knowledge. However, while the central nodal regions of the various disciplines of knowledge are neurally connected through their central neural regions the connection is not necessarily hierarchical.

With this structure of knowledge interdisciplinary research is possible since every part of any discipline can be selectively recalled, activated and used.

A New Era for Mathematics and Natural Science

With the internet both mathematics and natural science have descended from the esoteric Tower of Academe to the public domain. Starting with the innocent looking post on the math forum SciMath in 1997 regarding the equation 1 = 0.99... extensive spirited debate erupted and spilled over into hundreds of websites and their archives. Then the debate moved to FLT and the new real number system. It has eroded control of scientific information by the abstracting and indexing services and standard databases and brought it into the public domain. The immediate implications of these new developments are:

(1) Mathematics and natural science have become a material force for the elevation of the human condition in terms of practical applications, especially, technological development that can improve the quality of life.

(2) Error and ambiguity can no longer hide in the Tower of Academe for a long time as they are flushed out by global debate.

(3) Naturally, catastrophic events such as the disastrous final flight of the Columbia Space Shuttle due to theoretical, hence, ultimately mathematical error, can be minimized. The mathematical error in this case is the failure to device suitable physical theory to guide the program; such theory is forged by qualitative mathematics.

The disaster was one of the factors that raised standards in constructing a mathematical space.

The new era has matured for mathematics but it is still muted in natural science as there are many negative elements of traditions that must be cast away for natural science to advance.

Research and Development

Like mathematics scientific natural philosophy can no longer remain in the Tower of Academe. It must come down and become a material force for the elevation of the human condition. Since natural science is knowledge of how nature works it should provide the clue on how we can control nature to serve mankind. Research and development can provide the venue.

Research and development aims for sustainable economic-industrial development (which we simply refer to as development) of underdeveloped countries, *i.e.*, rapid generation and accumulation of enormous wealth, value or resources through production to improve the material conditions of society across the board, *i.e.*, positively affect all classes and sectors. Sustainable development means that it does not deprive future generations of the bounties of the Earth and the health of the ecology. The name of the game is technological advancement in production. Labor intensive production is inefficient and cannot generate enormous value. At the same time, production geared towards the domestic market is insufficient for who shall pay for the cost of devising or importing the hundreds of technology needed for development, particularly, production? The strategy for development we shall consider here suits underdeveloped countries alone because developed countries have the opposite problem: overproduction that compels factories to shut down and investments to plummet driving workers out of work during recession, its economic symptom.

It should be clear at the outset that the problem of development is distinct from the problem of social development that aims for equity, *i.e.*, equitable access to the resources and amenities of society, that economic-industrial development is a pre-condition for social development and social justice. Such social justice is not only unachievable when the resources are meager but also requires separate strategy for resolution. This clarification is quite important because development alone is bound to benefit only a small segment of society initially that tends to diminish enthusiasm for support from the other sectors of society, particularly, social scientists, but it is a pre-requisite for equitable distribution of resources.

Moreover, the possessors of such enormous resources in an underdeveloped country cannot consume but are bound to invest them and provide the wherewithal for setting up the intellectual, material and social infrastructure appropriate for developed society. Social infrastructure includes social services and benefits such as social security and unemployment benefits and adequate medical services. In general, income rises across the board in developed countries that even a janitor has a car and refrigerator. Still, access to basic amenities is unequal but this is no reason to oppose or be lukewarm toward development for it provides the basis for the resolution of the social problem of inequity aside from the fact that it sustains social stability (social instability is rooted in economic and political hopelessness). This should encourage social reformers to pursue parallel strategy for social development.

How will that strategy for development look like? Let us look at a bit of history. In the early fifties the International Monetary Fund and the World Bank promoted the Import Substitution Scheme to develop Asian countries under their influence. It sought to develop them from within by becoming self-reliant, especially, in technology. By the 1980s it was clear that the scheme was a total failure. Worse, the recipient countries remained not only underdeveloped but also buried deep in the quagmire of foreign indebtedness. The only exceptions were countries and territories that rejected the scheme and developed on their own. They are the newly industrializing countries (NIC) and territories of South Korea, Singapore, Taiwan and Hong Kong.

It is not surprising that an underdeveloped country cannot meet the requirements of development for who will pay for the cost of developing or importing the needed technologies? Now we understand why the developed countries today took centuries to get there. They did not have the resources to support development; nor did their colonies which could only provide cheap raw materials and labor and limited market.

Let us explore the horizons. There is enormous global resource, particularly, in developed countries that cannot be disposed locally or globally the reason for periodic recession. At the same time, recession or not, there are a myriad of things including technologies needed by developed countries year in and year out. How can an underdeveloped country exploit this situation? The answer is quite obvious: take a big chunk of that global resource and bring it to the country to provide for the cost of development including importation and generation of hundreds of needed technologies and setting up the appropriate intellectual, material and social infrastructure. How? By devising a couple of lynchpin technologies that can capture global monopoly during the 20-year period when their patents are in force. We call this approach Strategic Positioning which has a couple of important components aside from the lynchpin technologies. One is taking suitable diplomatic and trade initiatives to open up global markets and establish monopoly for the lynchpin technologies and, secondarily, identifying appropriate technologies for importation. Another is building up the appropriate intellectual and scientific infrastructure for developing appropriate scientific capability for the generation of needed technologies.

The lynchpin technologies cannot be found in high technology at this time because that is already monopolized by the developed countries and the NIC. We can only find them in the new epoch of GUT technology.

We conclude the subject with the observation that the economics being taught in developing and underdeveloped countries is focused on the dynamics of supply and demand which is appropriate for developed countries. This dynamics is missing in developing and underdeveloped countries, particularly, in Asia, Africa and Latin America where there is neither supply nor demand. The focus should be the study of the roots of backwardness and underdevelopment to pave the way for devising the strategy for development appropriate to the specific conditions of individual countries.

Philosophy and Status of Mathematics-Science Education

The current focus of mathematics education is computation and applications which is understandable since they meet the demands of the present methodology of science, quantitative modeling. This stress on computation and applications reflects itself in the emphasis on the unlearnable, *i.e.*, activities or concepts that do not challenge the mind such as computation and manipulation of symbols which are quickly learned in their areas of application even without a teacher. Moreover, such activities are quickly and more precisely done by machines. Furthermore, they give a distorted sense of what mathematics is. If we ask anyone what he thinks of mathematics he will invariably think of it as numbers and formulas. In actuality, mathematics as the primary language of science involves a myriad of other activity that we now collectively call qualitative mathematics. As part of reforms in mathematics education we need to introduce it at the initial phase of the training of the mathematician starting in high school. Currently, we are not producing mathematicians in number commensurate with the large number of mathematics majors who did well in college.

As we have seen earlier, computation and measurement, on the one hand, and qualitative mathematics, on the other, are complementary and indispensable to each other.

Another negative feature of current mathematics education is requiring too many examinations on the students that mainly test their memory, common sense and a bit of psychology. Even proofs of theorems can be memorized and a difficult problem can be solved in such exam provided the student anticipates it, solve problems of similar level of difficulty and memorize the reasoning in their solutions. Moreover, stuffing a mathematics course with too much material deprives the student more creative use of thought. In the classroom a single representative example that is thoroughly explained is more effective than a couple of examples that is explained superficially.

Here is something to ponder about: very few child prodigies in mathematics become mathematicians of note; so do wranglers of the four-century-old *Tripos* mathematics examinations in Cambridge University, UK. Among the few exceptions are Bertrand Russell (4th), L. C. Young (3rd) and John Littlewood, a senior wrangler (top notched). Moreover, it is not known if champions of the Mathematics Olympiad become mathematician. Mathematics is more than just memory, speed and ingenuity. It requires reflection and all the activity we have listed under qualitative mathematics but, most of all, creativity. If we do away with this unwanted burden on the mathematics student, his thought can be freed for challenging tasks and the discipline of mathematics will be more robust than it is today.

At the moment, nonlinear analysis is the only broad, rapidly expanding and exciting field of mathematics. One excitement here is the recent resolution of the 360-year-old Fermat's last theorem that spurred the development of the new real number system [17, 21] which opened up a number of fields not only in mathematics but also in natural science. The rest of mathematics is grinding to a halt, particularly, number theory, complex analysis and topology. Complex analysis has been given a boost recently with the introduction of the Complex Vector Plane that rectifies its problems and inadequacy (see Appendix to [19]). Group theory has been given a shot in the arm by cryptography, especially, as it applies to the problem of computer security. The current excitement in astronomy is the Hubble telescope that allows observation of the past and provides insights into our early universe. The next excitement, perhaps, is the recent discovery of the basic constituent of matter that led to the development of GUT.

Science education in all science disciplines is presently constrained by quantitative modeling. Therefore, as in mathematics, there is too much emphasis on computation, formulas and experimentation and less on knowing how nature works. In physics, for instance, there is need to understand such basic concepts as matter and energy which is only possible with qualitative mathematics. What makes the planet, star and galaxy spin? These are questions that the very young are curious about and there are no answers to them in quantitative modeling. They should be discussed starting in high school instead of stressing physical formulas. Physical concepts are more important and they generally do not require daunting computation to understand.

The good news in science is that there is generally great excitement in every natural science discipline. In biology, the excitement is in research on DNA (e.g., the Genome Project and its applications to medicine) and the stem cell. Then we can add the new knowledge about the gene. In chemistry, the excitement lies in its intersection with materials science. The excitement extends to the boundary of physics with psychology and the recent understanding of how the brain works. However, such excitement cannot be sustained for long unless it brings us to full knowledge of our universe and that is what scientific natural philosophy aims to fulfill. Fortunately, it has all it needs to achieve it.

To summarize, scientific natural philosophy provides a full view of nature based on the most updated scientific knowledge provided by GUT and its mathematics using the new methodology of qualitative-quantitative modeling.

Finally, we conclude this book by raising an important issue that has been hanging for quite some time: Is rational thought, the heart of natural science and the methodology of materialist philosophy, adequate to provide a comprehensive and full knowledge of nature within the domain of materialist philosophy? The answer is *yes* as we have seen in this journey that started with the resolution of Fermat's last theorem and the solution of the gravitational n-body problem in 1997. In particular, GUT has unified and explained the forces and interactions of nature from the atoms of quantum gravity through the cosmological vortices of macro gravity. It has also provided a complete cosmology of our universe from its birth at Cosmic Burst through its destiny as black holes back in dark matter.

Thoughts on Research

There is much wisdom in what this author's best academic friend, Professor V. Lakshmikantham, says on the subject of research: do it! Some researchers rely on the serendipity of science, others on sharpening their tools and reading the work of others leaving no time to think and ending up nowhere. Perhaps, this author's experience can offer some insights.

His academic career was interrupted 15 years straight during the political turmoil in the Philippines capped by the imposition of martial law by President Ferdinand Marcos, 1972 – 1986. The situation was the least this researcher needed at the time especially when he found himself not really in immediate danger but in a wide field in the general direction of the business end of the barrel of the gun. He left for the United States in 1971 and found a new imperative: build a support movement for the end of martial rule during which time he shelved his mathematical career completely.

When people's power ousted President Marcos in 1986 the author returned to the Philippines but faced a daunting challenge: he found himself at ground level while his international colleagues were up on the fifteenth floor of the Tower of Academe. Retracing their steps would have been impossible and historical precedence was not in his favor either. The closest to his experience was Bertrand Russell's when he was removed from his post as professor of mathematics at Cambridge University, England, for his pacifist activity during the First World War. When he got his post back 26 years later through the effort of British mathematician Paul Hardy, he went around mathematics and remained a philosopher.

Two years after his return to the Philippines the author was groping in the dark for a couple of years although he got results that would prove important later – the generalized integral and derivative. He knew that proving a theorem here and there would not help. The 15-year handicap needed something spectacular to overcome.

Then he posed the right question: why are there long-standing unsolved problems in mathematics and physics? The answer: the underlying fields are inadequate. The most famous problem in mathematics then was the 355-year-old Fermat's conjecture (popularly known as Fermat's last theorem or FLT) and in physics, the 200-year old gravitational n-body problem posed by Simon Marquis de Laplace at the turn of the 18th Century. He made a critique of the underlying fields of FLT – foundations, number theory and the real number system – and opened a can of worms that we saw earlier the most devastating of which being the inconsistency of the field axioms of the real number system that David Hilbert wanted to avoid. The inconsistency was first shown by Felix Brouwer by way of a counterexample to the trichotomy axiom [5] and, later, by this author with his own version of the counterexample to it [21,78]. The remedy was a simple construction of the decimals, using only three axioms, into the contradiction-free new real number system [21] that has, at the same time, countably infinite counterexamples to prove Fermat's conjecture false. It retains the valid and interesting properties of the real numbers and has new elements that qualitatively and quantitatively model some physical concepts such as the superstring.

The work was not over yet; no journal would touch the author's work with a ten–foot pole. Then he scoured mathematical landscape around the world to find the most influential mathematician in the field who can help. He found one: Professor V. Lakshmikantham, Editor-in-Chief of many scientific journals and founder of the broad and only rapidly expanding mathematical field today: nonlinear analysis. He was going to address the founding session of the International Conference on Dynamic Systems and Applications in Atlanta, 1995, organized by Professor M. Sambandham, Editor of the journal that published [21] and [44]. It was there where the author presented his introductory paper on the generalized integral and derivative, FLT, and the gravitational n-body problem, in the paper "Probabilistic mathematics and applications to dynamic systems including Fermat's last theorem" which was published in the proceedings [14]. Most of all, however, he got the lead to the World Congress of Nonlinear Analysts in Athens, 1996, sponsored by the International Federation of Nonlinear Analysts of which Professor Lakshmikantham was the President. He submitted his solution of the gravitational n-body problem which was accepted for presentation at the Congress. It was at this Congress where, through the initiative of Mrs. Escultura, Professor Lakshmikantham got a copy of his book, Diophantus: Introduction to Mathematical Philosophy (With probabilistic solution of Fermat's last theorem) [13], which he liked. He published the full paper, Exact solutions of Fermat's equation (A definitive resolution of Fermat's last theorem), in the journal, Nonlinear Studies, 1998, official publication of the International Federation of Nonlinear Analysts (now shared with the GVP – Lakshmikantham Institute for Advanced Studies, GVP College of Engineering, JNT University, India). One of the major results of the paper, characterization of undecidable propositions, gave the clue to the inadequacy of the present methodology of physics – mathematical modeling (now quantitative modeling). The introduction of qualitative mathematics and qualitative modeling as remedy was the crucial factor in the solution of the gravitational n-body problem that required the discovery of the superstring and the initial 11 natural laws needed for the solution that provided the foundations of GUT.

The passage of time has a way of distorting one's memory. In this author's earlier papers on GUT it is mentioned that qualitative modeling was introduced in 1997 to solve the gravitational n-body problem. In fact, it was introduced and was the main contribution of the author in his Ph. D. thesis [46] and in 1991 he published the book, Introduction to Qualitative Control Theory [22]. It was applied to physics for the first time in 1997 [41].

Perhaps success in research has something to do with one's academic training. This author's Ph. D. advisor, Professor L. C. Young of the University of Wisconsin lectured only on really new topics and did not give any examination. Instead, he threw this author into the lion's den, so to speak, to lecture at his faculty seminars (which were always well-attended) before mathematicians a number of whom really famous, some visiting professors. He discussed most of the hand written manuscripts of Young's book, Lectures on the Calculus of Variations and Optimal Control Theory [117] and recent papers by famous mathematicians then like L. Cesari and Peter Lax and reviewed mathematical papers submitted for publication (of course, Young signed the evaluation in each case this author being unknown). Young trained only 13 Ph.D.s who made a mark in diverse fields that fulfilled his vision: a good teacher is one who trains his pupil to beat him *somewhere* [118].

In his recommendation for the author's appointment at the University of Illinois Chicago which he got hold of many years later by virtue of the Freedom of Information Act, Professor Young wrote:

Dr. Edgar Escultura originally joined a remarkable group that I was fortunate enough to have working with me from about 1963 to 1970, a group consisting of post-doctoral visitors as well as really promising graduate students. I must admit that I had at first some doubts as to Escultura's suitability. In such a group, the standards are tough: only really good students as a rule will join it – students with the drive and confidence to tackle what is reputed to be hard. Much was expected, among other things, of the Seminar presentations: a tradition of excellence had established itself. Escultura surprised me by being able to fit into it. Indeed, on the strength of his Seminar presentations, I have no doubt that he can be an excellent teacher. My second surprise came when in due course Escultura went on to write a fine thesis under me. It was much more than a nice piece of work for solving some problem: there really were concepts of his own devising and the more I think of them, the more I feel that these concepts may prove important by opening up new methods in a number of fields far removed from that he was tackling. In this respect, the period of enforced interruption that he has experienced may be all to the good; it has allowed problems in a number of areas to mature to a stage at which they are ripe for his ideas. At any rate, I have no doubt that he will make significant contributions. I can recommend him in the highest terms.

The letter was quite providential. Unfortunately, he was unable to confirm it; although he reached the ripe old age of 95 and passed away on Christmas Eve, 2000, the author lost contact with him since 1993.

This author believes that everyone is endowed with the same potential. How much of that potential is realized depends on how he develops his thought through formal training, experience and his own initiatives. He agrees with Mahatma Gandhi that life is an experiment (the author extends this view to the development of biological species [31] where every species experiments with various anatomical and physiological innovations) and if one succeeds in an endeavor anyone else will. (Taken from "An Autobiography or The Story of My Experiments with Truth" by M. K. Gandhi, Navajivan Publishing House, Ahmedabad-14, 1927; translated from the Gujarati by Mahadev, 209 ed.)

The author stands on the shoulders of others. At the bottom of the article, Abstracts and Summary of Publications, on his websites, http://users.tpg.com.au/pidro/, http://edgareescultura.wordpress.com/, the author acknowledges his indebtedness to them:

I owe much of my achievements to others: R. A. Favila[a] and M. Sambandham[b] opened the doors to mathematics and science; L. C. Young[c] inspired and trained me well and V. Lakshmikantham[d] discovered and walked me over to the frontiers of science and mathematics.

[a]University of the Philippines, Quezon City, Philippines

[b]Morehouse College, Atlanta, USA

[c]University of Wisconsin, Madison, USA

[d]Florida Institute of Technology, Melbourne, USA

Glossaries

Basic Cosmic or Electromagnetic Waves. Waves generated by the normal vibration of the atomic nuclei that induces suitable synchronized vibration of dark matter (superstrings) that propagates across it.

Black Hole. Massive concentration of non-agitated superstrings in the eye of a cosmological vortex, e.g., galaxy or planet. Since black hole is dark matter, it does not suck matter. It is the eye of the cosmological vortex that nurtures and accumulates it that does; that suction is gravity.

Chaos. Indistinguishable mixture of orders or patterns of behavior. For example, at the onset of a hurricane where air molecules are rushing by the trillions towards tropical depression, one cannot monitor or predict the direction of motion of any molecule due to the immensity of their collisions. However, every molecule is subject to the laws of nature.

Cosmological Vortex. Vortex fluxes of superstrings in the Cosmos, e.g., planet, star, galaxy, a grain of cosmic dust.

Dark Matter. One of the two fundamental states of matter the other visible or ordinary matter. Dark matter consists manly of non-agitated but also semi-agitated superstrings and visible matter consists of agitated superstrings. The former comprises over 95% of our universe.

Energy. Energy is motion of matter. It follows that matter and energy are inseparable, i.e., neither pure energy nor pure matter exists. Of course, at this point this definition is partial since we do not know what matter is. We consider a couple of forms of energy from ordinary experience. Take heat which is kinetic or visible energy. It is due to the vibration of the atoms and molecules of an object. Its propagation through heat conduction is in accordance with the principle of resonance [40]. Another example is photon of light. Its kinetic and latent energy come from the motion of its toroidal flux [42]. The atom is quite familiar but we well define it later by the laws of nature.

Flux. Motion of matter whose location and direction are identifiable at each point, e.g., water current and basic cosmic wave. A vortex flux of superstrings is indirectly detected and measured by its charge which has direction at each point.

Mass. Measure of amount of matter in a physical system. In the metric system the basic unit of measure of mass is the kilogram and in the British system the pound.

Photon. A primum in flight, carried by basic cosmic wave, that has broken away from its loop. It is a rapid oscillation (planar) of its toroidal flux in the direction of flight.

Physical System. Any cluster, configuration or motion of superstrings, dark or visible. The superstring itself is a physical system. Wave as suitable synchronized motion of its medium is a physical system.

Primum. A primum is unit of visible matter, e.g., positron and electron. The electron and positive and negative quarks are basic prima for they comprise every light isotope of an atom. A simple primum is agitated helical bulge of a segment of a semi-agitated superstring [34,43,45]. Scooped up by basic cosmic wave it flattens as rapid oscillation along the direction of flight due to dark viscosity. When it breaks off from its loop it becomes a photon. While the primum is stable in flight being a loop the photon is stable only when its forward toroidal flux speed equals that of its carrier basic cosmic wave which is the speed of light of 3×10^{10} cm/sec; otherwise, it leaves the carrier and disintegrates, its toroidal flux remaining intact as (dark) non-agitated superstring. Thus, the photon has no rest mass; it disintegrates at rest.

Residual or Normal Vibration. Every atom vibrates (in fact, every piece of matter does) due to bombardment of its dark component by cosmic waves coming from all directions, its vibration characteristics determined by internal structure in accordance with the internal-external dichotomy law [42].

Superstring. Although the term *superstring* was used previously by others in the 1980s its identification as basic constituent of matter is fairly recent and the discovery of its configuration and properties even more recent. It is a closed helical loop like a lady's spring bracelet (Fig. 3.1). It contains a superstring called toroidal flux traveling through its helical cycles at the staggering linear speed of 7×10^{22} cm/sec which is 10^{12} times the speed of light. This is a constant of nature and uniform for all fluxes of superstrings including electric current in accordance with the synchronization principle of the energy conservation equivalence law Chapter 3. The toroidal flux has a toroidal flux which is a superstring, etc., ad infinitum. It is this configuration called generalized nested fractal sequence that makes the superstring indestructible. A superstring is agitated if its cycle length CL is less than 10^{-16} m (meters), semi-agitated if $10^{-16} < CL < 10^{-14}$ m and agitated if $CL > 10^{-14}$.

The Big Bang. Explosion of a black hole; that black holes was destiny of the core of a previous universe. It created a **super…super depression** in dark matter that evolved into a **super…super galaxy**, our universe.

The Cosmic Sphere. Expanding wave front caused by the Big Bang; its explosion, the **Cosmic Burst**, that occurred at t = 1.5 billion years after the start of the Big Bang released the compressed semi-agitated superstrings trapped between its inner and outer boundaries. They formed pure prima called **quasars** during the initially hot post-Cosmic Burst that peaked at t = 2.5 billion years after the start of the Big Bang [104]. They were the predecessor of the galaxies. The Cosmic Burst marks the first appearance of visible matter and, therefore, the birth of our universe.

Turbulence. Coherent or stable flux such as typhoon or water wave and current.

Wave. Wave in general is due to suitably synchronized motion of the medium. Where does its energy come from? It comes from the motion of the generator, e.g., wind motion, tectonic plate subduction at suitable distance from the ocean surface creating tsunami. It is reinforced by the induced synchronized vibration of the medium via resonance. In a sense, a wave is an illusion for no mass travels with it. A wave is like a "traveling" neon light; it does not really travel its apparent motion being due to synchronized switching on and off. Consider a spherical stone thrown into a pool. It pushes a cylindrical column of water downward. Water pressure pushes it back upward beyond the surface due to momentum while being pulled back by gravity until it grinds to a halt at its crest (highest point) and reverses downwards due to gravity beyond the water surface due to momentum. Then water pressure pushes it back upward again and the cycle is repeated. How do we account for its sinusoidal profile? It is due to water viscosity at the cylindrical boundary of the column that exerts a retarding pull on the rising column and gives it that profile. The vibration or oscillation of the water column resonates with the surrounding molecules of water and induces similar sinusoidal motion of concentric cylindrical columns of water propagating outward. Ocean waves are generated mainly by wind thrust oblique to the ocean surface that creates micro vibration of the water molecules (called micro-component of turbulence [35]) sustained and reinforced by continued wind flow and gravity until large waves form and travel in the direction of forward resultant of wind flow. They break when they become too large and heavy to sustain.

References

[1] Aguda, B. In: *The cell cycle* (abstract), Program of the 3rd World Congress of Nonlinear Analysts, July **2000**, Catania, Italy p. 59.

[2] Angay, M. C. *On the Symmetries and Probability Distributions of Oscillations, Generalized Limit and Derivative, and Applications*, Master's Thesis, **1996**, University of the Philippines Los Baños, College, Laguna, Philippines.

[3] Astronomy (a) Auugust **1995**, (b) January **2001**, (c) June **2002**.

[4] Atsukovsky, V.A. *General ether-dynamics; simulation of the matter structures and fields on the basis of the ideas about the gas-like ether*; Energoatomizdat: Moscow, **1990**.

[5] Benacerraf, P. and Putnam, H. *Philosophy of Mathematics*, Cambridge University Press: Cambridge, **1985**.

[6] Chandra Prakash, K, Sarma, V. S. R. S. and Escultura, E. E., (a) *Economic-Industrial Development for Sustainable Development of Underdeveloped Countries*, Proc. International Conference on Sustainable Development, Excel India Publishers (ISBN: 978-93-80043-76-0), Vol. *1*, Chennai, Feb. 1 – 6, **2010**, 241 – 243; (b) *Building Green to Attain Sustainability*, Proc. International Conference on Emerging Trends in Engineering, Dr. J. J. Magdum College of Engineering, Jaysingpur, Kolhapur Dist., Maharashtra, India (ISBN 978-81-910130-0-9), Feb. 19 - 22, **2010**, pp. pp. 171 – 176; (c) *Building Green - Practices and Challenges*, Proc. , International Conference on Environmental Sustainability with Green Building Technology, Chennai, India Meenakshi Sundararajan Engineering college, Chennai, India (ICESGBT-10), March 15 – 17, **2010**, pp. 192 - 198.

[7] Corporate Mathematical Society of Japan. *Encyclopedic Dictionary of Mathematics* (Ito, K., Editor), MIT Press (2nd ed): Cambridge, MA, **1993**.

[8] Davies, P. C. and Brown, J. *Superstring: A Theory of Everything?* Cambridge University Press: New York, **1988**.

[9] Davies, P. J. and Hersch, R. *The Mathematical Experience*, Birkhäuser: Boston, **1981**.

[10] Discover, (a) Feb. **1996**, (b) July **1999**, (c) Nov. **1998**, (d) Dec. **1999**.

[11] Edgar, G. A. *Measure, Topology and Fractal Geometry*, Springer-Verlag: New York, **1990**.

[12] Encarta Premium, **2008**.

[13] Escultura, E. E., *Diophantus: Introduction to Mathematical Philosophy (With probabilistic solution of Fermat's last theorem)*, Kalikasan Press: Manila, **1993**.

[14] Escultura, E. E. *Probabilistic mathematics and applications to dynamic systems including Fermat's last theorem*, Proc. 2nd International Conference on Dynamic Systems and Applications: Atlanta, May 27 – 31, **1999**, pp. 147 – 152.

[15] Escultura, E. E. Bhaskar, T. G. Leela, S., Laksmikantham, V., Revisiting the hybrid real number system. *J. Nonlinear Analysis, C-Series: Hybrid Systems*, May **2009**, *3*(2), pp. 101 – 107.

[16] Escultura, E. E. Extending the reach of computation. *J. Applied Mathematics Letters*, **2008**, *21*(10), pp. 1074 – 1081.

[17] Escultura, E. E. Exact solutions of Fermat's equation (A definitive resolution of Fermat's last theorem. *J. Nonlinear Studies*, **1998**, *5*(2), pp. 227 - 254.

[18] Escultura, E. E. *Recent verification and applications*, Proc. 2rd International Conference on Tools for Mathematical Modeling, St. Petersburg, **1999**, *4*, pp. 116 - 29.

[19] Escultura, E. E. The generalized integral as dual of Schwarz distribution, invited paper. *J. Nonlinear Studies*.

[20] Escultura, E. E. Set-valued differential equations and applications to quantum gravity. *J. Mathematical Research*, **2000**, *6*, St. Petersburg, pp. 221 - 224.

[21] Escultura, E. E. The new real number system and discrete computation and calculus. *J. Neural, Parallel and Scientific Computations*, **2009**, *17*, pp. 59 – 84.

[22] Escultura, E. E. *Introduction to Qualitative Control Theory*, Kalikasan Press: Manila, **1991**.

[23] Escultura, E. E. The new mathematics and physics, *J. Applied Mathematics and Computation*. **2003**, *138*(1), 145 – 169.

[24] Escultura, E. E. Chaos, turbulence and fractal. *Indian J. Pure and Applied Mathematics*, **2001**, *32*(10), pp. 1539 – 1551.

[25] Escultura, E. E. The mathematics of the grand unified theory, Proc. 5th World Congress of Nonlinear Analysts. *J. Nonlinear Analysis, A-Series: Theory: Method and Applications*, **2009**, *71*, pp. e420 – e431.

[26] Escultura, E. E. The mathematics of the new physics. *J. Applied Mathematics and Computations*, **2002**, *130*(1), pp. 149 - 169.

[27] Escultura, E. E. Dynamic Modeling and the new mathematics and physics. *J. Neural, Parallel and Scientific Computations* (NPSC), **2007**, *15*(4), PP. 527 – 538.

[28] Escultura, E. E. The theory of intelligence and evolution. *Indian J. Pure and Applied Math.*, **2003**, *33*(1), PP. 111 – 129.

[29] Escultura, E. E. The physics of the mind, accepted. *J. Science of Healing Outcomes*.

[30] Escultura, E. E. The origin and evolution of biological species. *J. Science of Healing Outcomes*, **2010**, *6-7*, pp. 17 - 27.

[31] Escultura, E. E. The Unified Theory of Evolution, invited. *International Journal of Biological Sciences and Engineering.*

[32] Escultura, E. E. Genetic Alteration, Modification, Sterilization With Applications to the Treatment of Genetic Diseases, accepted. *J. Science of Healing Outcomes.*

[33] Escultura, E. E. Superstring loop dynamics and applications to astronomy and biology. *J. Nonlinear Analysis, A-Series: Theory: Methods and Applications,* **1999**, *35*(8), pp. 259 – 285.

[34] Escultura, E. E., From macro to quantum gravity. *J. Problems of Nonlinear Analysis in Eng'g Systems,* **2001**, *7*(1), pp. 56 – 78.

[35] Escultura, E. E. Turbulence: theory, verification and applications. *J. Nonlinear Analysis, A-Series: Theory, Methods and Applications,* **2001**, *47*(8), pp. 5955 – 5966.

[36] Escultura, E. E. Vortex Interactions. *J. Problems of Nonlinear Analysis in Engineering Systems,* **2001**, *7*(2), pp. 30 – 44.

[37] Escultura, E. E. Dynamic Modeling of Chaos and Turbulence, Proc. 3rd World Congress of Nonlinear Analysts. *J. Nonlinear Analysis, A-Series: Theory: Method and Applications,* **2005**, *63, 5-7,* pp. e519-e532.

[38] Escultura, E. E. Global geology and oceanography, invited. *International J. Earth Sciences and Engineering.*

[39] Escultura, E. E. *Dynamic Modeling and Applications*, Proc. 3rd International Conference on Tools for Mathematical Modeling, 7, State Technical University of St. Petersburg, St. Petersburg, **2003**, pp. 103 - 114.

[40] Escultura, E. E. The Pillars of the new physics and some updates. *J. Nonlinear Studies,* **2007**, *14(3)*, pp. 241 – 260.

[41] Escultura, E. E. The solution of the gravitational n-body problem. *J. Nonlinear Analysis, A-Series: Theory, Methods and Applications,* **1997**, *30*(8), pp. 5021 - 5032.

[42] Escultura, E. E. The grand unified theory, contribution to the Felicitation Volume on the occasion of the 85th birth anniversary of Prof. V. Lakshmikantham. *J. Nonlinear Analysis, A-Series: Theory: Method and Applications,* **2008**, *69*(3), pp. 823 – 831.

[43] Escultura, E. E. Qualitative model of the atom, its components and origin in the early universe, Proc. 5th World Congress of Nonlinear Analysts. *J. Nonlinear Analysis, B-Series: Real World Applications,* **2009**, *11*, pp. 29 – 38.

[44] Escultura, E. E. *Dynamic and mathematical models in physics*, Proc. 5th International Conference on Dynamic Systems and Applications, June 30 – July 5, **2007**, Atlanta.

[45] Escultura, E. E. The basic concepts and dynamics of quantum gravity with applications, invited. *J. Nonlinear Studies.*

[46] Escultura, E. E. The trajectories, reachable set, minimal levels and chain of trajectories of a control system, Ph. D. thesis, University of Wisconsin, **1970**.

[47] Escultura, E. E. *Qualitative modeling for complex systems*, Proc. 23rd EURO Conference, July 5 – 8, **2009**, Bon, Germany.

[48] Falconer, K. J. *The Geometry of Fractal Sets*, Cambridge University Press, Cambridge, **1986**.

[49] Feix, M. Rapport d'activite du 01 Juillet 1990 au 01 Juillet, Centre National De la Recherches Sci. (*Phys. Math., Modélisation et simulation*), Lyons, **1992**.

[50] Fleming, W. H.; Young, L. C. Representation of generalized surfaces as mixtures. *Rens. Cir. Mat.* **1956**, *5*(2), pp. 117 – 144.

[51] Gale Encyclopedia, **2000** ed.

[52] Gerlovin, I. L. *The Foundations of United Theory of Interactions in a Substance*, Energoattomizdat: Leningrad, **1990**.

[53] Gleick, J. Chaos: *Making a New Science*, New York, **1988**.

[54] Gödel, K. On formally undecidable propositions of Principia Mathematica and related systems. *Monatshefte für Mathematik und Physik,* **1931**, Springer-Verlag, *38*, 173 – 198.

[55] Gödel, K. On undecidable propositions of formal mathematical systems. *Lectures,* Institute for Advanced Study, Princeton, **1934**.

[56] Gödel, K. On intuitionistic arithmetic and number theory (translation). *Ergebnisse eines mathematischen Kollquiums,* **1933**, *4*, pp. 34 - 38.

[57] Gödel, K. On the length of proofs. *Ergebnisse eines mathematischen Kollquiums,* **1933**, *7*, 23 – 38 (translation).

[58] Gödel, K. Remarks before the Princeton bicentennial conference on problems in mathematics, **1946**.

[59] Gödel, K. An unsolvable problem of elementary number theory, presented to the American Mathematical Society, April 19, **1935**.

[60] Gödel, K. The completeness of the axioms of the functional calculus of logic, a rewritten version of Gödel's Ph.D. dissertation at the University of Vienna, **1930**.

[61] Gödel, K. Some mathematical results on completeness and consistency, On formally undecidable propositions of Principia mathematica and related systems I, and On completeness and consistency; main paper received by the Vienna Academy of Sciences for publication on 17 November **1930**; an abstract was presented to the Academy by Hans Hahn, October 23, **1930**.

[62] Gudkov, V. V. Existence of a traveling wave solution for multi-component systems. *J. Differential Equations,* **1995**, *31*(1), pp. 153 – 162.

[63] Gudkov, V. V. The explicit form of the wave solutions of the evolutionary equations. *J. Comp. Math. Phys.,* **1996**, *36*(3), pp. 135 – 340.

[64] Gudkov, V. V. Helical solutions for a model of electron transport in molecular systems. *Proc. Latvian Academy of Sciences,* **1998**, Section B, 52, No. *5*(598), pp. 248 – 250.

[65] Gudkov, V. V. *Matrix sol. of diffusion equation*, Proc. 3rd World Congress of Nonlinear Analysis, Catania, **2000**. *J. Nonlinear Analysis, A-Series: Theory, Methods and Applications,* **2001**, *47*(3), pp.161 – 164.

[66] Gudkov, V. V. Torus as a geometrical image of a family of matrix solutions of the nonlinear Klein-Gordon equation. *J. Nonlinear Analysis, A-Series: Theory, Methods and Applications,* **2003**, *48*, pp. 255 – 262.

[67] Gudkov, V. V.; Escultura, E. E. Mathematical models on the way from superstring to photon. *J. Nonlinear Analysis, B-Series: Real World Applications,* **2002**, *3*, pp. 375 – 382.

[68] Gudkov, V. V.; Escultura, E. E. The Helix and Other Optimal Configurations of Matter and Applications. *J. Nonlinear Analysis and Phenomena,* **2006**, *3*(2), pp. 46 - 64.

[69] Guinness Book of Records, **2002**.

[70] Hawking, S. *A Brief History of Time*, Bantam Press: London, **1989**.

[71] Hocking, J. G.; Young, G. S. *Topology*, Addison-Wesley Publishing Company: Reading, Massachussets, **1961**.

[72] Horgan, H. The death of proof. *Scientific American*, **1993**, *5*, pp. 74 – 82.

[73] Jesudason, C. G. Some consequences of an analysis of the Kelvin-Clausius entropy formulation based on traditional axiomatic. *Entropy*, **2003**, *5*, pp. 252 – 270.

[74] Kaku, M and Thompson, J. *Beyond Einstein*, Anchor Books: New York, **1995**.

[75] Kline, M. Mathematics: The Loss of Certainty. *Oxford University Press*, New York, **1980**.

[76] Krylov, B. V.; Shehegelov; B. F. *Mathematical methods in the physiology of sensory Systems*, 2nd International Conference on Tools for Mathematical Modeling, St. Petersburg, Russia, **1999**.

[77] Lakatos, I. *Proofs and Refutations* (J. Worral and E. Zahar, eds.), Cambridge University Press: Cambridge, **1976**.

[78] Lakshmikantham, V.; Escultura, E. E.; Leela, S. The Hybrid Grand Unified Theory, Atlantis (World Scientific): Paris, March **2009**.

[79] Lakshmikantham, V. *The Origin of Human Past: Children of Immortal Bliss*, *Bharatiya*, Vidya Bhavan: Bombay, **1999**.

[80] Lee, P. Y.; Chew, T. S. *Integration of highly oscillatory functions in the plane*, Proc., Asian Math. Conference, Hong Kong, **1990**.

[81] Merchbacher, E. *Quantum Mechanics* (2nd ed): John Wiley & Sons, Inc. New York, **1970**.

[82] New Scientist, (a) *Cannibalism by giant galaxies*, July **1997**; (b) *Fractal universe*, November **1999**.

[83] Nieper, H. A. *Revolution in Technology, Medicine and Society*, Management Interessengemeinschaft für Tachyon-Feld-Energie: Odenburg, FRG, *1*, **1984**.

[84] Osipenko; G. S.; Pokrovsky, A. N.; Krylov, B. V.; Plakhova, V. B.; *Math. Modeling of pain sensation*, Proc. 2nd International Conference on Math Modeling, St. Petersburg, Russia, **2000**.

[85] *Our Solar System*, A Reader's Digest Young Families Book. *Joshua Morris Publishing*, Inc., Westport, CT, **1998**.

[86] Pendick D., Fires in the Sky. *Earth*, June **1996**, *20*, pp. 62 – 64.

[87] Pickover, C. A. *Chaos and Fractals*, Elsevier Science: Amsterdam, **1998**.

[88] Pokrovsky, A. N. *Small parameters in neuron models*, Proc. 2nd International Conference on Tools for Mathematical Modeling, St. Petersburg, Russia, **1989**.

[89] Pokrovsky, A. N. *On potentiation of the synapses*, *Biofizika*, **1989**, *34*(400), p. 7.

[90] Pokrovsky, A. N. Mathematical model of the neural network memorizing the time interval. *Biofizika*, **1996**, *41*(5), pp. 1102 – 1101.

[91] Pontrjagin, L. S.; Boltyanskii; V. G., Gamkrelidze; R. V.; Mischenko, E. F. *The Mathematical Theory of Optimal Processes*, (K. N. Trirogoff, Tran.; L. W. Neustadt, Ed.), Interscience Publishers: New York, **1962**.

[92] Ridpath, I. *Atlas of Stars and Planets*, George Philip Ltd.: London, **1999**.

[93] Royden, H. L. *Real Analysis*, MacMillan, 3rd ed.: New York, **1983**.

[94] Science, *Ultraenergetic particles*, Aug. **1998**, *281*(5379), pp. 241 – 242.

[95] Science, *Watching the universe's second biggest bang*, March **1999**, *283*(5410), pp. 2013 – 2014.

[96] Science, *Glow reveals early star nurseries*, July **1998**, pp. 332 – 333.

[97] Science, *Galaxy's oldest stars shed light on Big Bang* (hydrogen, helium lithium and others no heavier than boron), Nov. **1998**, *282*(5382), pp. 1428 – 1429.

[98] Science, (a) *Cosmic motion revealed*, December **1998**, **282**, **1998**,*5397*; (b) *No Backing off from accelerating universe*, Nov. **1998**, *282*(5392), pp. 1248 – 1249.

[99] Science, (a) *Weighing in on neutrino mass*, **1997**, **280** pp. 1689 – 1691; (b) *Neutrinos weigh in*, Dec. **1998**, *282*(5397), pp. 2158 – 2159; (c) *Search for neutrino mass is a big stretch for three labs*, February **1999**, *283*(5405), pp. 928 – 929.

[100] Science, New clues to the habits of heavy weights (black holes at craters of galaxies). **1999**, *283*(5401), pp. 480 – 481.

[101] Science, Pluto: *The planet that never was*, Jan. **1999**, *283*(5399), p. 157.

[102] Science, (a) *Starbirth, gamma blast hint at active early Universe*, Dec., **1998**, *282*(5395), p. 1806; (b) *Gamma burst promises celestial reprise*, Jan. **1999**, *283*(5402), p. 616; (c) Powerful cosmic rays tied to far off galaxies (100 million times reached in largest particle accelerators), Nov. **1998**, *282*(5391), p. 1023.

[103] Science, The mystery of the migrating galaxy clusters, Jan. **1999**, *283*(5402), pp. 625 – 626.

[104] Scientific American, (a) *The quasars and early galaxies*, April **1983**, pp. 708 – 745.

[105] Scientific American, *Galactic collision*, April **1995**, pp. 11 – 14.

[106] Smoot, G. and Davidson, K. *Wrinkles in Time*, Avon Books: New York, **1988**.

[107] (a*) The Earth Atlas*; (b) *The Oceans Atlas*, Dorling Kindersley: London, **1994**.

[108] *The Great Discoveries*, Time Books, **2001**.

[109] Trippi, R. R. *Chaos & Nonlinear Dynamics, Financial Markets*, Irwin: New Your, **1995**.

[110] Wiles, A. Modular Elliptic curves and Fermat's last theorem. *Annals of Mathematics*, **1995**, *141*(3), pp. 443 – 551.

[111] Wiles, A. and Taylor, R. Ring-theoretic properties of certain Hecke algebras, *Ann. Math.*, **1995**, *141*(3), pp. 553 – 572.

[112] Young, L. C. Approximations by polygons. *Proc. Roy. Soc. A*, **1933**, *141*, pp. 325 – 341.

[113] Young, L. C. *Generalized curves and the existence of an attained absolute minimum in the calculus of variations*, Camp. Rend. Soc. Sci. Letter, Cl III, Varsovie, **1937**, *30*, pp. 211 – 234.

[114] Young, L. C. Generalized surfaces in the calculus of variations: I Generalized Lipschitzian surfaces, *Annals of Mathematics*, June **1942**, *43*(2), pp. 84 – 103. II. Mean surfaces and the theory of the problem ∫∫f(x,y,p,q)dxdy = Min., *Annals of Math.*, **1942**, *43*(3), pp. 530 – 544.

[115] Young, L. C. Some applications of the Dirichlet integral to the theory of surfaces. *Trans. Amer. Math. Soc.*, **1948**, *64*, pp. 317 – 335.

[116] Young, L. C. *On Generalized surfaces of finite topological types*, Memoirs of the AMS: Philadelphia, **1955**.

[117] Young, L. C., *Lectures on the Calculus of Variations and Optimal Control Theory*, W. B. Saunders: Philadelphia, **1969**.

[118] Young, L. C. *Mathematicians and Their Times*, North-Holland: Amsterdam, **1980**.

[119] Young, L. C. and Nowosad, P. Generalized curve approach to elementary particles. *JOTA*, **1983**, *41*(1), pp. 261 – 273.

[120] Zeigler, B.P *An Introduction to Calculus" Course Based on DEVS: Implications of a Discrete Reformulation of Mathematical Continuity, presented* at the *International Conference on Simulation in Education ICSiE'05*, New Orleans, Jan. 23 – 25, **2005**.

[121] Zeigler, B.P. *Continuity and Change (Activity) are Fundamentally Related in DEVS Simulation of Continuous Systems*, Keynote Talk at AI, Simulation and Planning AIS'04, Korea, Oct. 4-6, **2004**.

Index

A

Adjacent decimals 38, 39
affine transformation 25, 142
agitated, semi-agitated and non-agitated superstring
Andromeda 90
angular and linear momentum 66
anthropic principle 152
Archimedean property 5, 39, 149
asteroid 94
Asteroid Belts 88
atmospheric Turbulence 135
atom 70
atrophy 118, 122
axiom of choice 5, 149

B

background radiation 107, 152
balls of fire and earthlight 76
Banach-Tarski paradox 5
barber paradox 5, 12
basic cosmic or electromagnetic wave 66
basic integers 32
basic prima 64
Bertrand Russell 5
Big Bang 152
Birthmark 125
geological and atmospheric sciences 108
biology 108
black hole 93, 96, 105
Bose-Einstein condensate 74
Brain wave 109
brittle and malleable material 77
Brouwer, L. E. J. 6
Brownian motion 78

C

canonical Euler equations 24
Cantor's diagonal method 35, 150
cardinality 35
central neural region 112
chaos 16, 24, 25, 130
charge 64
chattering control 14
Clausius, R. 61
COBE Project 107, 152
cognition 112
Columbia Space Shuttle 63, 81, 152
Comet 98, 141
completeness axiom 5
complex systems 8